CONCEPTS OF GENETICS

Breno Lisboa

Toronto Academic Press

CONCEPTS OF GENETICS

Breno Lisboa

Toronto Academic Press
4164 Lakeshore Road
Burlington ON L7L 1A4
Canada
www.tap-books.com

Email: orders@arclereducation.com

ISBN: 978-1-77956-297-5

ABOUT THE AUTHOR

Breno Lisboa is a committed bioinformatician with a strong expertise in bioinformatics, biotechnology, synthetic biology and genetic engineering. Breno is now studying for a Doctorate in Bioinformatics at the Federal University of Minas Gerais (UFMG) in Belo Horizonte, Brazil. His research focuses on investigating the origins of life and the significance of microscopic terrestrial algae in the development of Earth's soils and the evolution of land plants. His study integrates evolutionary biology and ecology to provide new insights into these basic processes. Breno acquired expertise in the field of biotechnology via his Master's program at the Federal University of Bahia (UFBA). During his studies, he specifically concentrated on conducting structural analyses of house dust mites. His research efforts have made significant contributions to the advancement of hypoallergenic vaccinations and enhanced diagnostic techniques. He obtained a Bachelor's degree in Biotechnology from UFBA and conducted research at the Technical University of Munich. During his research, he used structural bioinformatics to investigate the immunological cross-reactivity between mite allergens and helminth proteins. Additionally, he achieved a gold medal in the iGEM 2022 competition, showcasing his inventive methodology and unwavering commitment to scientific superiority. Breno Lisboa's fervor for bioinformatics, together with his varied experiences and unwavering commitment to furthering scientific knowledge, establishes him as a proficient and inventive researcher in the realm of bioinformatics.

Table of Contents

List of Figures

List of Tables

PREFACE

The field of human genetics provides insight into the fundamental nature of our biological identity. It demonstrates how the complex genetic code in our DNA influences not only our observable traits but also our vulnerability to illnesses and our reactions to the environment. Genetics plays a crucial role in all aspects of our lives, including our response to environmental toxins and the benefits our bodies gain from exercise. The genetic composition of our bodies is a continually changing phenomenon that interacts with our actions and surroundings. The intricate interaction shown here emphasizes the crucial significance of understanding human genetics, especially in the field of healthcare. By comprehending these core ideas, we improve our capacity to collect and use genetic and genomic data efficiently, resulting in more accurate diagnoses and customized therapies. Significantly, humans possess a DNA sequence that is 99.9% identical, providing strong evidence of our shared genetic heritage. However, it is the remaining 0.1% of our genetic makeup that is responsible for the vast array of human variability. This diversity presents itself in two main forms: mutations, which are infrequent and often associated with illnesses, and polymorphisms, which are more prevalent and have diverse effects on health. Single Nucleotide Polymorphisms (SNPs) are especially intriguing since they involve the alteration of a single unit of DNA. The single nucleotide polymorphisms (SNPs) are widely distributed across our genome, and each person has millions of them. It is essential to comprehend that not all genetic differences result in illness. Certain factors may have no noticeable impact on health, while others may affect qualities as basic as blood type. The link between genetic variation and health is nuanced, hindered by our inadequate understanding of the exact genes implicated in certain illnesses and the various ways genes interact with one another and with environmental and behavioral variables. Technological advancements have significantly improved our capacity to detect genetic variants. Currently, we have the capability to analyze many genes at the same time, identifying changes with exceptional speed and precision. This advancement is crucial for enhancing our comprehension of human genetics and its ramifications for health and disease. As we explore the captivating realm of human genetics, we embark on a voyage that has the potential to transform our approach to health, medicine, and our understanding of the distinctive characteristics that define us as individuals. The following sections will navigate you through this intricate but enthralling discipline, establishing the basis for a more profound understanding of the significance genetics has in our existence.

– Author

CHAPTER 1

Molecular Foundations of Genetics

LEARNING OBJECTIVES

After studying this chapter, you will be able to:

- Describe the structure and function of DNA and RNA
- Discuss the DNA replication
- Give the overview central dogma
- Explain the gene and mechanism of gene action

INTRODUCTORY EXAMPLE

Molecular Genetics Is the Study of the Molecular Structure of DNA:

Molecular genetics, a branch of biology that explores the structure and function of genes at the molecular level, is a captivating field with profound implications for understanding life itself. By delving into the intricate mechanisms governing heredity, development, and disease, molecular genetics elucidates the fundamental building blocks of organisms – DNA and its associated molecules.

At the heart of molecular genetics lies DNA, or deoxyribonucleic acid, a remarkable molecule that encodes the genetic instructions essential for life. Composed of nucleotide units, DNA forms the iconic double helix structure, wherein two complementary strands wind around each other. This elegant architecture not only imparts stability but also facilitates the faithful transmission of genetic information from one generation to the next through the process of replication.

Central to the functioning of DNA is its ability to store and transmit genetic information in the form of genes. Genes are specific sequences of nucleotides that serve as the blueprint for synthesizing proteins – the molecular machines responsible for carrying out virtually every function in living organisms. Through the intricate process of gene expression, information encoded within DNA is transcribed into messenger RNA (mRNA) and subsequently translated into proteins, orchestrating the myriad biological processes that underpin life.

The elucidation of the structure of DNA by James Watson and Francis Crick in 1953 heralded a new era in biology, paving the way for groundbreaking discoveries in molecular genetics. One such discovery was the understanding of the genetic code – the correspondence between the sequence of nucleotides in DNA and the sequence of amino acids in proteins. This universal code, deciphered through the collaborative efforts of scientists worldwide, provided profound insights into the molecular basis of life and revolutionized fields ranging from medicine to biotechnology.

Moreover, molecular genetics has unraveled the intricate regulatory mechanisms that govern gene expression, shedding light on how cells precisely control which genes are turned on or off in response to internal and external cues. From the intricate interplay of transcription factors and enhancers to the dynamic modifications of chromatin structure, these regulatory mechanisms ensure the precise orchestration of cellular processes essential for development, homeostasis, and adaptation.

Furthermore, molecular genetics has played a pivotal role in advancing our understanding of genetic diseases – disorders caused by mutations in DNA that disrupt normal gene function. Through the identification and characterization of disease-causing genes, researchers have made significant strides in diagnosing, treating, and even preventing genetic disorders, offering hope to millions affected by conditions once deemed incurable.

UNIT INTRODUCTION

In biology, the field of molecular genetics is devoted to the study of genes and their molecular makeup, functions, and roles in inheritance. Fundamentally, molecular genetics explores the complex processes that underlie the transfer of genetic information from one generation to the next and how organisms use this information to control their growth, development, and general functioning. The gene, which is a particular DNA nucleotide sequence that encodes instructions for the synthesis of functional RNA molecules or proteins, is one of the basic building blocks of molecular genetics. Deciphering the intricacies of genetic inheritance and trait expression requires an understanding of the structure and regulation of genes.

The understanding of the genetic code, which determines how nucleotide sequences are translated into the amino acid sequences of proteins, is one of the ground-breaking discoveries made possible by advances in molecular genetics. Researchers can now analyze and manipulate DNA with previously unheard-of precision and efficiency thanks to techniques like polymerase chain reaction (PCR) and DNA sequencing, which have completely changed the field.

Molecular genetics is also essential to many other fields of biotechnology and biological research. It offers opportunities for the development of diagnostic instruments, treatments, and preventive measures by shedding light on the molecular causes of diseases. By modifying DNA sequences specifically, genetic engineering techniques—such as CRISPR-Cas9 gene editing technologies—allow scientists to better understand gene function and create organisms with desired traits.

Molecular genetics helps us understand biodiversity and evolutionary processes by revealing the genetic mechanisms underlying speciation and adaptation, in addition to its biomedical applications. Through the analysis of molecular profiles across various species, scientists are able to deduce evolutionary relationships and track the genesis of genetic variation.

All things considered, the field of molecular genetics is still vibrant and fast developing, with broad ramifications for conservation, agriculture, medicine, and other areas. Its multidisciplinary approach, which incorporates concepts from molecular biology, biochemistry, and genetics, guarantees its ongoing importance in expanding our knowledge of life at the molecular level.

1.1. STRUCTURE AND FUNCTION OF DNA AND RNA

Learning Objectives

- Describe the DNA structure and genetic code
- Explain the DNA types
- Define the DNA function

1.1.1. DNA Structure and Genetic Code

Deoxyribonucleic acid, or DNA, is the genetic material found in humans and nearly all other living things. Adenine (A), guanine (G), cytosine (C), and thymine (T) are the four chemical bases that make up the code containing the information in DNA. Approximately 3 billion bases make up human DNA, and over 99% of those bases are shared by every individual. Similar to how letters in the alphabet appear in a certain order to form words and sentences, the sequence, or order, of these bases determines the information available for constructing and maintaining an organism.

Base pairs are units of DNA that are formed when two bases pair up, such as A with T and C with G. Additionally, each base is joined to a phosphate and sugar molecule. A nucleotide is made up of a base, a sugar, and a phosphate. A double helix is a spiral made up of two long strands of nucleotides. The base pairs in the double helix structure resemble the rungs of a ladder, while the vertical side pieces of the ladder are made up of sugar and phosphate molecules.

The ability of DNA to replicate, or create duplicates of itself, is a crucial characteristic. The base sequence can be duplicated using a pattern found on each strand of DNA in the double helix. This is crucial during cell division because every newly formed cell requires an identical replica of the DNA found in the parent cell (Figure 1.1).

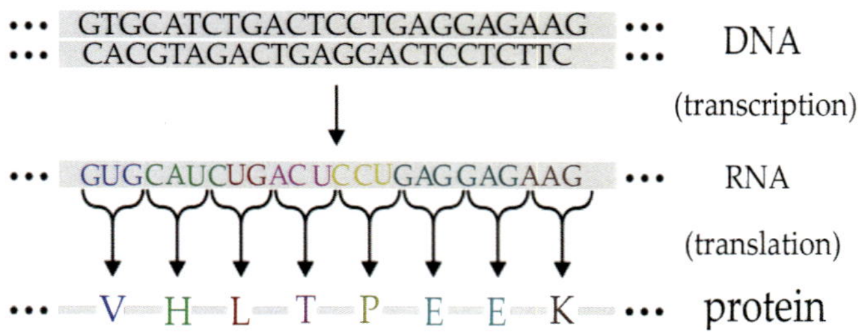

Figure 1.1. *DNA and the genetic code.*

Source: https://en.m.wikipedia.org/wiki/File:Genetic_code.svg.

1.1.1.1. What Is DNA?

"Hereditary materials, or genetic instructions, are carried and transmitted from parents to offspring by a group of molecules called DNA."

This also holds true for viruses, since the genetic material of most of these entities is made up of either DNA or RNA. For example, the genetic material of some viruses might be RNA, whereas the genetic material of other viruses might be DNA. The RNA that makes up the Human Immunodeficiency Virus (HIV) is transformed into DNA once it has bonded to the host cell. In addition to ensuring that all living things pass on their genetic information, DNA is essential for the synthesis of proteins. The DNA found in each nucleus of a eukaryotic organism is known as nuclear DNA. It codes for most of the genomes of the organisms; the remainder is handled by the DNA of plastids and mitochondria.

Keyword

Genetic Information – This in bacteria and many viruses is encoded in DNA, but some viruses use RNA.

Mitochondrial DNA is the name given to the DNA found inside the cell's mitochondria. It is passed down from the mother to her offspring. The human mitochondrial DNA contains about 16,000 base pairs. In a similar vein, plastids are vital to photosynthesis and have their own DNA.

1.1.1.2. Full-Form of DNA

Deoxyribonucleic acid is the full form given to DNA. This organic substance has a distinct molecular structure. It is present in every eukaryotic and prokaryotic cell (Figure 1.2).

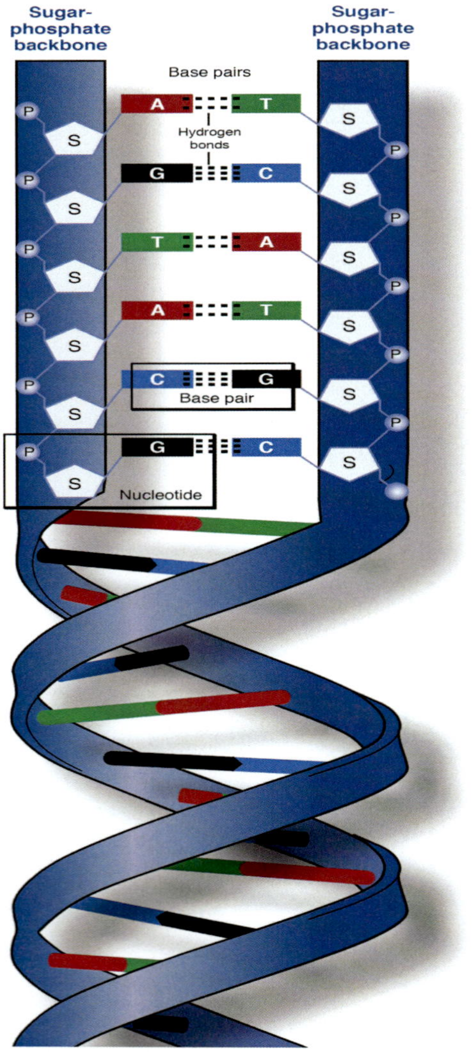

Figure 1.2. *Deoxyribonucleic acid (DNA).*

Source: By National Human Genome Research Institute – https://www.genome.gov/genetics-glossary/Phosphate-Backbone, Public Domain, https://commons.wikimedia.org/w/index.php?curid=125057969.

1.1.2. DNA Types

There are three different DNA types:

1. **A-DNA:** It is a double helix with a right hand that resembles the B-DNA form. Dehydrated DNA adopts an A form that shields it from harsh environments like desiccation. Additionally, DNA loses its solvent when bound to proteins, and it assumes the A form.
2. **B-DNA:** This is the most prevalent form of the right-handed helix found in DNA. Under normal physiological conditions, most DNA has a B-type conformation.
3. **Z-DNA:** It is a left-handed DNA whose double helix winds in a zigzag fashion to the left. Alexander Rich and Andres Wang made this discovery. It is thought to have some function in gene regulation because it is located before a gene's start site.

1.1.3. DNA Function

DNA is the genetic material that carries all the hereditary information. Genes are small segments of DNA, consisting mostly of 250 to 2 million base pairs. A gene codes for a polypeptide molecule, where a sequence of three nitrogenous bases stands for one amino acid.

To create distinct proteins, polypeptide chains are further folded into secondary, tertiary, and quaternary structures. Every organism's DNA contains a large number of genes, which allow for the formation of various protein types. In most organisms, proteins serve as the primary structural and functional molecules.

Apart from storing genetic information, DNA is involved in:

- Replication process: Transferring the genetic information from one cell to its daughters and from one generation to the next and equal distribution of DNA during the cell division.

- Mutations: the changes that occur in the DNA sequences.
- Transcription.
- Cellular metabolism.
- DNA fingerprinting.
- Gene therapy.

1.1.3.1. Why DNA Is Called a Polynucleotide Molecule?

Deoxyadenylate (A), deoxyguanylate (G), deoxycytidylate (C), and deoxythymidylate (T) are the nucleotides that make up the DNA molecule and when combined, form long chains known as polynucleotides. This is why DNA is referred to as a polynucleotide. The two polynucleotide chains constitute DNA are described by the structure of DNA.

Did you get it?

1. **What is DNA?**
2. **Why DNA Is called a polynucleotide molecule?**

1.2. DNA REPLICATION

Learning Objectives

- **Explain the Errors in DNA Replication:** Each double strand of DNA splits into two single strands during each cell division. A new complementary DNA strand can be created using any one of these single strands as a template. Every new cell has a complete genome of its own as a result. We call this process DNA replication. DNA polymerase enzymes direct replication, which is governed by the Watson-Crick pairing of the template strand's bases with incoming deoxynucleoside triphosphates. It is a multi-enzyme process that's complicated, especially in eukaryotes. Figure 1.3 describes a simple version of bacterial DNA replication.

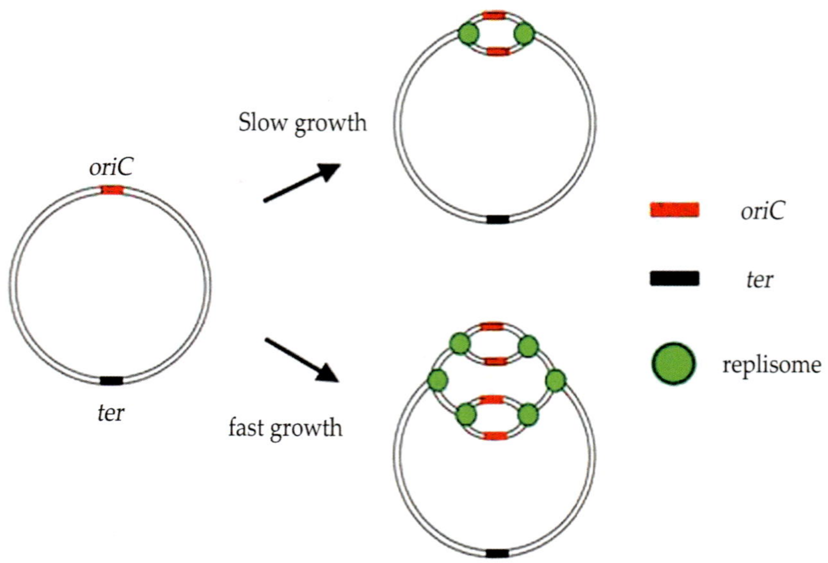

Figure 1.3. *Simplified representation of DNA replication in bacteria.*

Source: https://www.frontiersin.org/files/Articles/428011/fmicb-09-02819-HTML/image_m/fmicb-09-02819-g001.jpg.

While the lagging strand template becomes available, the opposing strand is copied in brief bursts of roughly 1000 bases while the leading strand is synthesized continuously. After their discoverers, Reiji and Tsuneko Okazaki, the resulting short strands are referred to as Okazaki fragments. At least three different DNA polymerases are found in bacteria: Pol I, Pol II, and Pol III. Pol III is primarily responsible for chain elongation. It is odd that DNA polymerases need a brief primer with a free 3′-hydroxyl group in order to start DNA synthesis from scratch. This is created in the lagging strand by an

RNA polymerase known as DNA primase, which can create a brief RNA molecule with a length of about 20 bases by using the DNA template. After that, Pol III can take over, but along the way, one of the previously created short RNA fragments gets in the way.

Now that the RNA has been broken down, Pol I steps in and uses its 5′-to 3′-exonuclease activity to fill the space with DNA until it reaches a continuous stretch of DNA. This creates a space between the 5′ end of the DNA that Pol III previously synthesized and the 3′ end of the newly synthesized DNA. The enzyme DNA ligase, which forms a covalent bond between a 5′-phosphate and a 3′-hydroxyl group, fills the gap. More specialized texts go into greater detail about the more complicated process of DNA replication beginning at the leading strand.

Keyword

Synthesis – It is the production of chemical compounds by reaction from simpler materials.

1.2.1. Errors in DNA Replication

The replication of DNA is not flawless. In DNA replication, mistakes arise when the wrong base gets incorporated into the expanding DNA strand. Mismatched base pairs, or mispairs, are the result of this. DNA repair enzymes have developed to fix these errors, and DNA polymerases have the ability to proofread. Mispairs sometimes make it through and are added to the genome during the subsequent replication cycle. These mutations could have no effect, cause the organism to die, result in a genetic illness or cancer, or give the organism an advantage over its neighbors, allowing for evolution through natural selection.

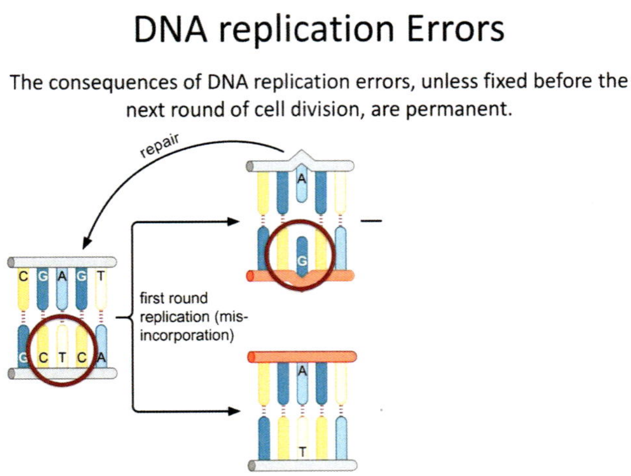

Figure 1.4. *Errors in DNA replication.*

Source: https://slideplayer.com/slide/13398358/80/images/2/DNA+replication+Errors.jpg.

1.3. THE CENTRAL DOGMA

Learning Objectives

- Focus on transcription
- Explain the translation
- Discuss the genetic code
- Describe the transfer RNA
- Explain the wobble hypothesis

The 'Central Dogma' is the process by which the instructions in DNA are converted into a functional product. It was first proposed in 1958 by Francis Crick, discoverer of the structure of DNA.

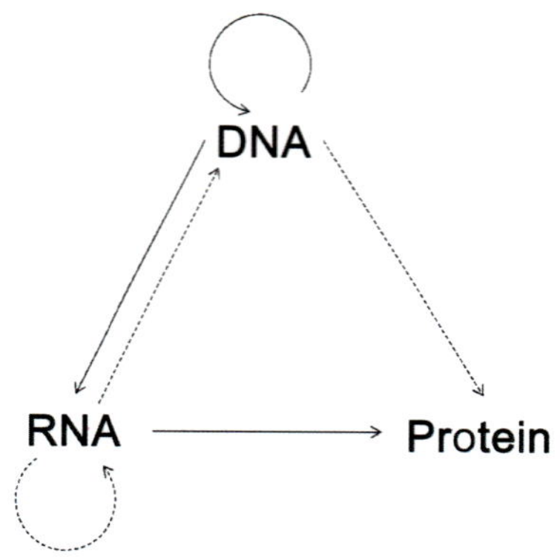

Figure 1.5. *The flow of genetic information in cells from DNA to messenger RNA (mRNA) to protein.*

Source: Kubo, A., Teti, R., Ullah, A. S., Iwadate, K., & Segreto, T., (2021). Determining surface topography of a dressed grinding wheel using bio-inspired DNA-based computing. Materials, 14(8), 1899. https://doi.org/10.3390/ma14081899.

The transmission of genetic information from DNA to messenger RNA (mRNA) to protein in cells is described by the fundamental tenets of molecular biology. It says the sequence of mRNA molecules, which determines the sequence of proteins, is specified by genes. The cell protects its DNA and copies it into RNA because the information contained in DNA is essential to cellular function. Each nucleotide that an enzyme reads from the

DNA strand is increased by one on the mRNA strand. It takes more work to translate this information into a protein because one amino acid in a polypeptide sequence is represented by three mRNA nucleotides (Figure 1.5).

1.3.1. Transcription

The process of copying DNA to mRNA, which contains the instructions required for protein synthesis, is known as transcription. There are two main stages to transcription. RNA polymerase enzymes are first involved in the formation of pre-messenger RNA. The process is dependent on Watson-Crick base pairing, and the single strand of RNA that is produced is the DNA sequence's reverse complement. RNA splicing is the process by which the pre-messenger RNA is "edited" to yield the target mRNA molecule.

1.3.1.1. Transcription: DNA to RNA

The process of converting a DNA sequence into a complementary RNA copy is called transcription. RNA polymerase reads a DNA sequence during transcription, creating an antiparallel, complementary RNA strand. The initial stage of gene expression is transcription. A transcript is a segment of DNA that has been transcribed into an RNA molecule. While some transcripts encode one or more proteins, others are employed as structural or regulatory RNAs.

- **Remember:** Messenger RNA (mRNA), which is produced during transcription and utilized during translation to produce the desired protein if the transcribed gene encodes one, is the outcome of transcription.

1.3.1.2. Formation of Pre-Messenger RNA

There are similarities between the transcription and DNA replication mechanisms. Similar to DNA replication, transcription cannot happen unless the double helix is partially unwound. The RNA polymerase enzymes catalyze this process. Only one strand is transcribed, as opposed to the two strands that are copied during DNA replication. The complementary strand is known as the antisense strand, and the strand containing the gene is known as the sense strand. The sense strand is duplicated in the mRNA that is created during transcription, but the antisense strand is the one that is transcribed.

NTPs, or nucleoside triphosphates, align with Watson-Crick base pairing along the antisense DNA strand. To create a pre-messenger RNA molecule complementary to a section of the antisense DNA strand, RNA polymerase assembles the ribonucleotides. When the RNA polymerase enzyme encounters a triplet of bases, which is interpreted as a "stop" signal, transcription is terminated. To form the double helix again, the DNA molecule rewinds (Figure 1.6).

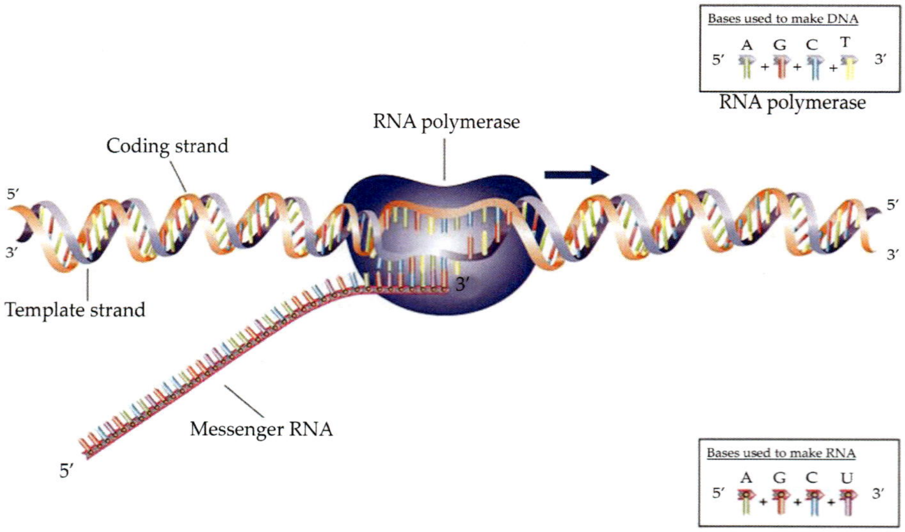

Figure 1.6. *Transcription simplified representation of the formation of pre-messenger RNA (orange) from double-stranded DNA (blue) in transcription.*

Source: By Genomics Education Program – Process of transcription, CC BY 2.0, https://commons.wikimedia.org/w/index.php?curid=50542917.

1.3.1.3. RNA Splicing

Thus produced pre-messenger RNA has introns that are not needed for the synthesis of proteins. RNA splicing is the process of chopping up the pre-messenger RNA to cut out the introns and produce messenger RNA (mRNA) (Figure 1.7).

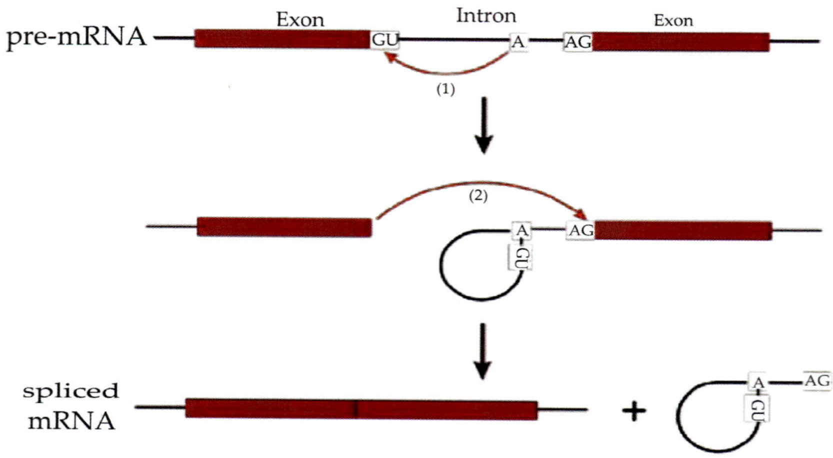

Figure 1.7. *Introns are spliced from the pre-messenger RNA to give messenger RNA (mRNA).*

Source: By BCSteve – Own work, CC BY-SA 3.0, https://commons.wikimedia.org/w/index.php?curid=30096313.

1.3.1.4. Alternative Splicing

Individual exons are either spliced or included during alternative splicing, resulting in a variety of potential mRNA products. Every mRNA product encodes a distinct protein isoform, and because of differences in their peptide sequences, these protein isoforms exhibit different biological activities. Up to 60% of human gene products are thought to go through alternative splicing. There are several known mechanisms of alternative splicing; Figure 1.8 depicts the approach of alternative splicing.

Remember

It should come as no surprise that aberrant splicing patterns can result in various disease states, such as cancer.

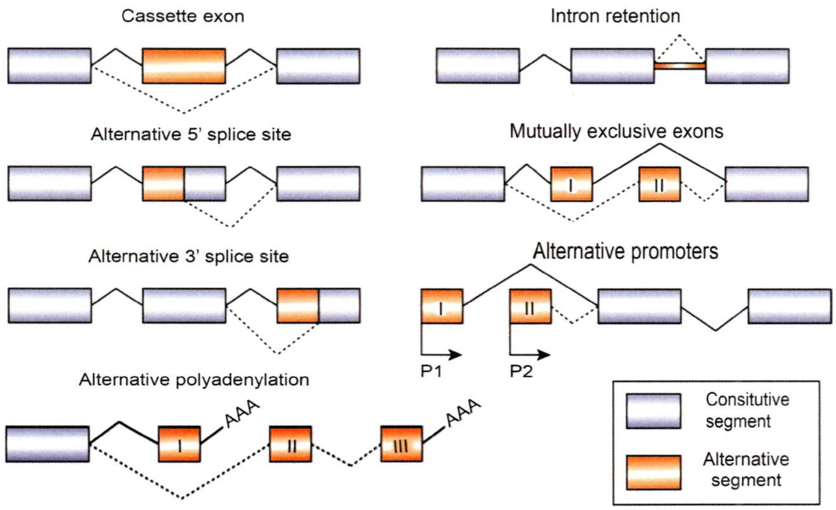

Figure 1.8. *Alternative splicing: Several different mechanisms of alternative splicing exist – a cassette exon can be either included in or excluded from the final RNA, or two cassette exons may be mutually exclusive.*

Source: Liu, Q., Fang, L., & Wu, C., (2022). Alternative splicing and isoforms: From mechanisms to diseases. Genes, 13(3), 401. https://doi.org/10.3390/genes13030401.

Keyword

Genetic regulation is the process of turning genes on and off to control when, where, and how much they are expressed.

A single gene transcript (RNA) can have thousands of different splicing patterns, and as a result, code for thousands of different proteins. This means that a diverse proteome can be produced from a relatively small genome. Alternative splicing contributes to protein diversity. Genetic regulation depends on splicing because protein expression is altered when the splicing pattern is altered in response to cellular circumstances.

1.3.1.5. Reverse Transcription

Reverse transcription is the process of converting RNA into DNA. Through this process, reverse transcriptase enzymes facilitate the use of RNA as genetic material by retroviruses, such as the human immunodeficiency virus (HIV). In biotechnology, reverse transcriptase enzymes are also useful because they enable the conversion of RNA to DNA for methods like PCR.

1.3.2. Translation

The mRNA produced during transcription is carried by the cell's synthesis of proteins, the ribosome, from the nucleus into the cytoplasm.

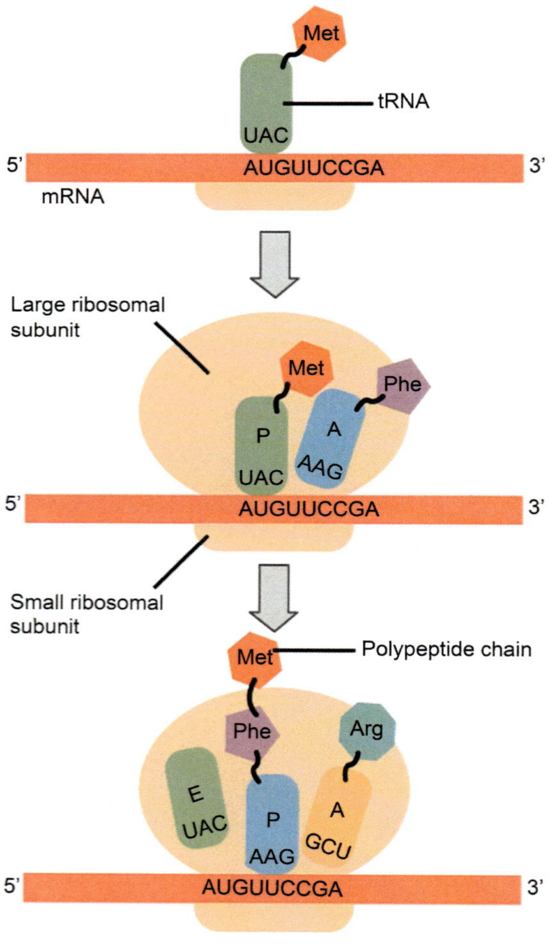

Figure 1.9. *Translation.*

Source: https://courses.lumenlearning.com/suny-biology1/chapter/ribosomes-and-protein-synthesis/.

It controls the synthesis of proteins here. Transfer RNA (tRNA) is necessary for protein synthesis; messenger RNA is not directly involved in this process. Translation is the process by which tRNA helps mRNA control the synthesis of proteins.

An enormous complex of RNA and protein molecules makes up the ribosome. A codon is a three-base stretch of mRNA (triplet) that carries the information for a single amino acid. Each codon engages in Watson-Crick base pairing with the anticodon of a distinct transfer RNA (tRNA) molecule as the mRNA travels through the ribosome. At its 3′-terminus, this tRNA molecule contains an amino acid that is integrated into the expanding protein chain. After that, the tRNA leaves the ribosome. The steps involved in protein synthesis are depicted in Figure 1.9.

1.3.2.1. Translation: RNA to Protein

The process of decoding and translating mRNA into a polypeptide sequence—also referred to as a protein—is called translation. The mRNA directs this process of protein synthesis, which is carried out with the assistance of a ribosome, a sizable complex of proteins and ribosomal RNAs (rRNAs). During translation, a cell assembles the new polypeptide chain by decoding the genetic message contained in the mRNA.

Translating the codon sequence on the mRNA strand is known as transfer RNA or tRNA. A free amino acid is transferred by tRNA from the cytoplasm to a ribosome, where it attaches itself to the expanding polypeptide chain. A stop codon on the mRNA is reached by tRNAs as they keep adding amino acids to the polypeptide chain's growing end. Next, the completed protein is released into the cell by the ribosome.

1.3.3. The Genetic Code

The 20-letter code of amino acids, which are the building blocks of proteins, is translated from the four-letter code of DNA according to a set of rules known as the genetic code. The codons, which are groups of three nucleotide combinations that make up the genetic code, each represent a distinct amino acid or stop signal. Francis Crick and associates initially explained the idea of codons in 1961. Marshall Nirenberg and Heinrich Matthaei conducted experiments that year that marked the beginning of the decipherment of the genetic code. It was demonstrated that the amino acid phenylalanine was specifically coded for by the RNA sequence UUU. Nirenberg, Philip Leder, and Gobind Khorana fully characterized every three-letter codon and its corresponding amino acid after making this discovery. They also identified the remaining portions of the genetic code (Figure 1.10).

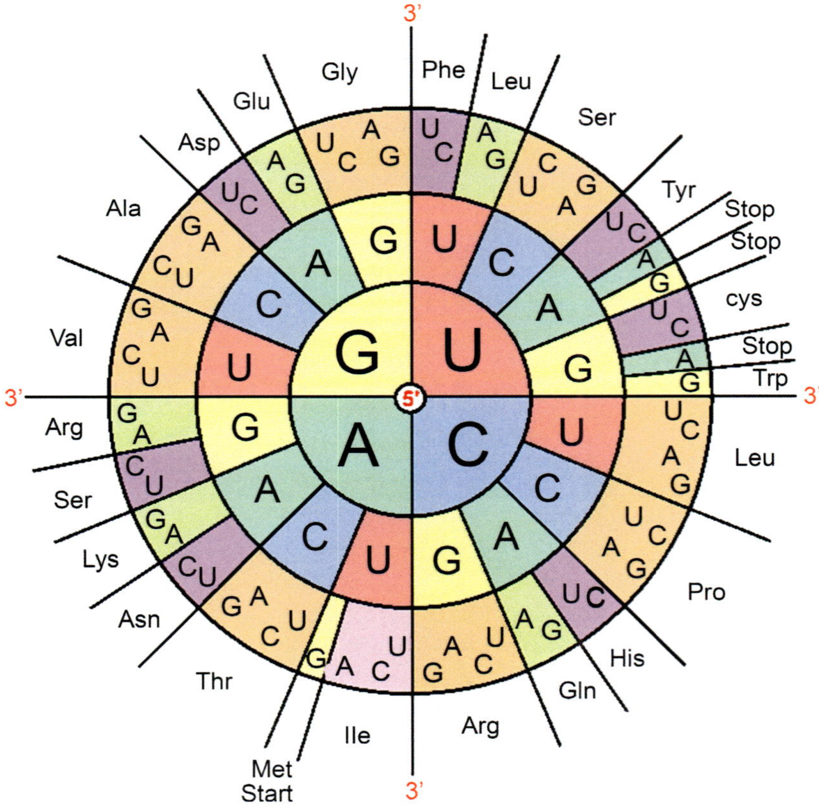

Figure 1.10. *Understanding the genetic code.*

Source: By Mouagip – Codons aminoacids table.png, Public Domain, https://commons.wikimedia.org/w/index.php?curid=5986132.

From the four nucleotides, there are 64 possible permutations, or combinations, of three-letter nucleotide sequences. Three of these 64 codons are stop signals, and the remaining 61 represent amino acids. The genetic code is called degenerate, or redundant, because multiple codons may code for the same amino acid, even though each codon is specific for only one amino acid (or one stop signal). The fact that the genetic code does not overlap and that each nucleotide is only a part of one codon—a single nucleotide cannot be a part of two adjacent codons—is also very important to note. Moreover, the genetic code is almost universal, with very few documented variations. For example, mitochondria have a different genetic code that differs slightly.

Keyword

A nucleotide is a molecule that is the basic building block of DNA and RNA, the nucleic acids that carry genetic information inside cells.

1.3.3.1. Genetic Code Table

A genetic code is the entirety of the relationships between codons and amino acids; this code is frequently represented in Table 1.1.

Table 1.1. Genetic Code

First letter	Second letter: U	Second letter: C	Second letter: A	Second letter: G	Third letter
U	UUU Phe	UCU Ser	UAU Tyr	UGU Cys	U
U	UUC Phe	UCC Ser	UAC Tyr	UGC Cys	C
U	UUA Leu	UCA Ser	**UAA Stop**	**UGA Stop**	A
U	UUG Leu	UCG Ser	**UAG Stop**	UGG Trp	G
C	CUU Leu	CCU Pro	CAU His	CGU Arg	U
C	CUC Leu	CCC Pro	CAC His	CGC Arg	C
C	CUA Leu	CCA Pro	CAA Gln	CGA Arg	A
C	CUG Leu	CCG Pro	CAG Gln	CGG Arg	G
A	AUU Ile	ACU Thr	AAU Asn	AGU Ser	U
A	AUC Ile	ACC Thr	AAC Asn	AGC Ser	C
A	AUA Ile	ACA Thr	AAA Lys	AGA Arg	A
A	**AUG Met**	ACG Thr	AAG Lys	AGG Arg	G
G	GUU Val	GCU Ala	GAU Asp	GGU Gly	U
G	GUC Val	GCC Ala	GAC Asp	GGC Gly	C
G	GUA Val	GCA Ala	GAA Glu	GGA Gly	A
G	GUG Val	GCG Ala	GAG Glu	GGG Gly	G

Source: https://courses.lumenlearning.com/wm-nmbiology1/chapter/reading-codons/It is evident that multiple codons display multiple amino acids in the table. Leucine, for instance, can be written in six different ways in the mRNA language.

The universality of the genetic code is one of its main features. This suggests that, with very few exceptions, almost every species uses the genetic code to synthesize proteins. To put it another way, genetic code is the base nucleotide sequence on DNA that is translated into the amino acid sequence of the protein that has to be synthesized.

1.3.3.2. Properties of Genetic Code

- Triplet code;
- Non-ambiguous and universal;
- Degenerate code;
- Nonoverlapping code;

- Commaless;
- Start and stop codons;
- Polarity.

These properties of genetic code are explained below.

1. **Triplet Code:** A set of bases that designates an amino acid is known as a codon or code word. Strong evidence exists to support the idea that a protein's amino acid is encoded for a triple nucleotide sequence e.g., a triplet is the code. The first nucleotide's four bases. 3 base codons are created using i.e., (A, G, C, and U). Sense codons, which code for amino acids constitute the 64 codons. Because multiple codons for a single amino acid indicate that more than one codon for that amino acid exists, there are 64 codons for 20 amino acids.
2. **Commaless Code:** There is no space for punctuation between them, indicating that each codon is next to its neighbor without a single nucleotide separating them.
3. **Nonoverlapping Code:** A nucleotide that joins one triplet never joins the next triplet because the code is read in groups of three sequentially.

For example:

- 5′-UCU-3′ codes for serine;
- 5′-AUG-3′ codes for methionine.

4. **Polarity:** Every triplet is read from 5′ ⊠ 3,' with 5′ as the starting base, followed by the middle base and 3′ as the final base. This suggests that codons are fixed polarity, meaning that if a codon is read the other way around, its base sequence would reverse and specify two distinct proteins.
5. **Degenerate Code:** All amino acids are coded by different codons, with the exception of tryptophan (UGG) and methionine (AUG) i.e., a small number of codons have synonyms; this feature of the genetic code is called degeneracy. For instance, in yeast mitochondria, UGA codes for tryptophan.
6. **Start and Stop Codons:** The initiating or start codon is typically the AUG codon. This can be in prokaryotes (N-formylmethionine) or eukaryotes (methionine) to start the polypeptide chain.

   Conversely, UAG, UAA, and UGA are referred to as stop codons or termination codons. They never code for any amino acids and are not read by any tRNA molecules.
7. **Non-Ambiguous and Universal:** A given codon will only code for a single amino acid because the genetic code is unambiguous. Furthermore, it is observed that every organism has the same genetic code e.g., they apply to everyone.

1.3.3.3. Exceptions to the Code

Since most genes in microbes and plants have similar START and STOP signals and assign similar codons to identical amino acids, the genetic code is universal. There are, however, a few known exceptions, the majority of which involve giving an amino acid one or both of the STOP codons. Apart from this, even though GUG is intended to code for valine, both AUG and GUG may code for methionine as a starting codon. This violates the non-ambiguousness property. It follows that a small number of codes frequently deviate from the universal or unambiguous code.

1.3.4. Transfer RNA

The tertiary structure of transfer RNA is well defined and typically depicted in two dimensions as a cloverleaf shape. Figure 1.11 provides further details on the structure of tRNA.

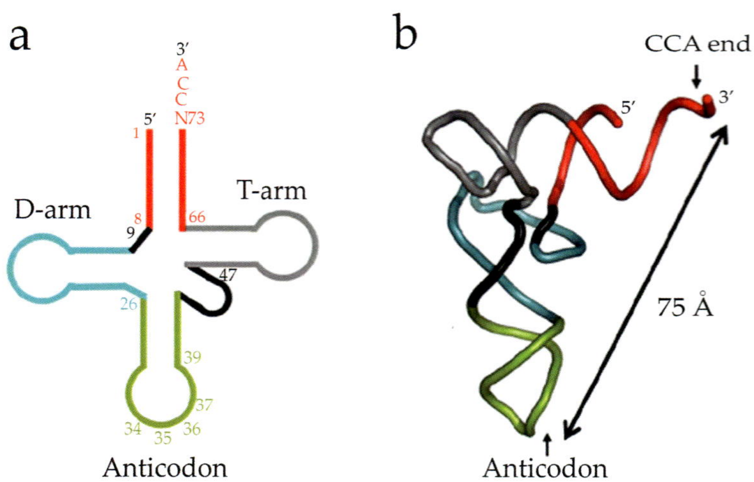

Figure 1.11. *Secondary (a) and tertiary (b) structures of transfer RNA (tRNA). The cloverleaf secondary structure folds to the L-shaped tertiary structure; both structures have corresponding colors. tRNA has the universal single-stranded CCA-3′ terminus and a neighboring nucleotide at position 73 called "discriminator," which is also non-base-paired in most cases.*

Source: Tamura, K., (2015). Origins and early evolution of the tRNA molecule. Life, 5(4), 1687–1699. https://doi.org/10.3390/life5041687.

There is a unique tRNA (or set of tRNAs) for every amino acid. For instance, the histidine (tRNAHis) and phenylalanine (tRNAPhe) tRNAs are distinct from one another. During protein synthesis, every amino acid is linked to its tRNA by the 3′-OH group to create an ester, which then combines with the terminal amino acid's ⊠-amino group to form a new amide bond, or peptide bond (Figure 1.12). Esters and amines generally react favorably, but the rate of reaction is significantly faster in the ribosome.

Aminoacyl tRNA synthetase, an enzyme, recognizes the distinct tertiary structure of every transfer RNA molecule and adds the appropriate amino acid to the 3′-end of the uncharged tRNA. Modified nucleosides play a crucial role in maintaining the stability of the tRNA structure. A few of these modifications are displayed in Figure 1.13.

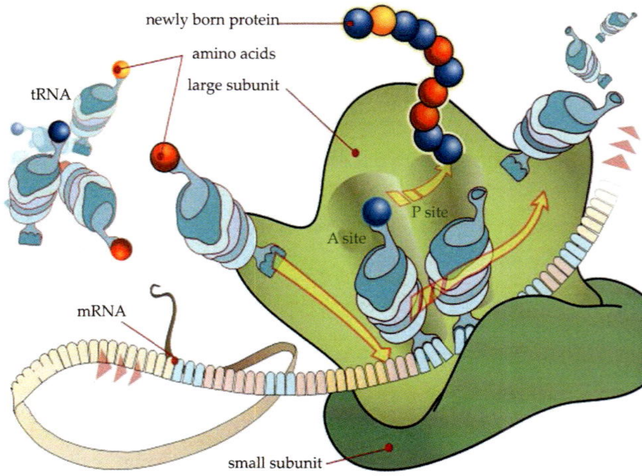

Figure 1.12. *Protein synthesis, reaction of the growing polypeptide chain with the 3′-end of the charged tRNA. The amino acid is transferred from the tRNA molecule to the protein.*

Source: https://courses.lumenlearning.com/wm-nmbiology1/chapter/reading-steps-of-translation/.

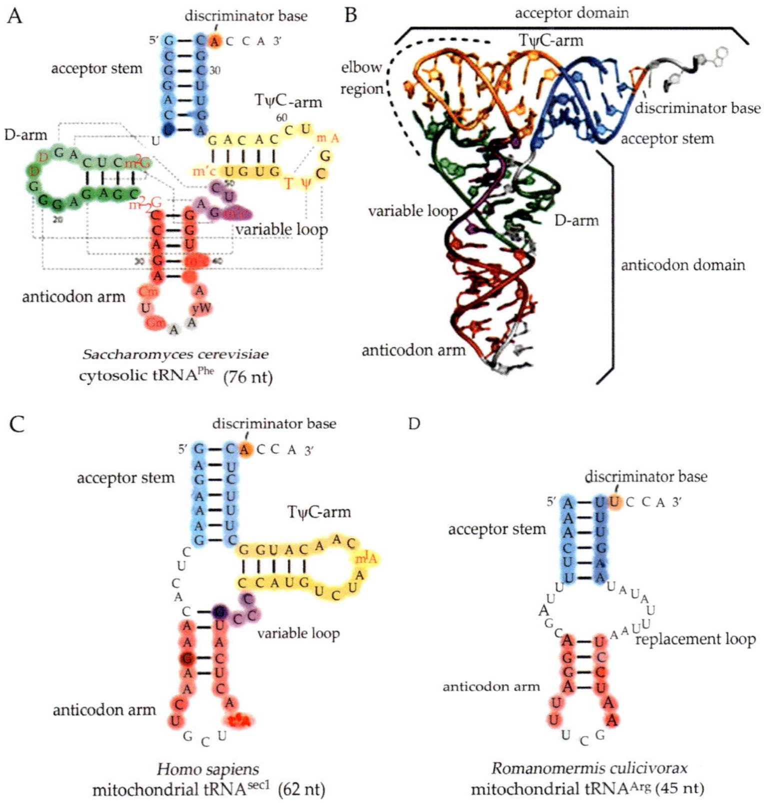

Figure 1.13. *Modifications in tRNA structures.*

Source: Lorenz, C., Lünse, C. E., & Mörl, M., (2017). tRNA modifications: Impact on structure and thermal adaptation. Biomolecules, 7(2), 35. https://doi.org/10.3390/biom7020035.

1.3.5. The Wobble Hypothesis

Upon close examination, it becomes evident that the variability in codons is most prominent in the third position when it comes to a specific amino acid (e.g., alanine has codons GCU, GCC, GCA, and GCG). Crick and Brenner proposed that a singular tRNA molecule can detect codons with differing bases at the 3′-end due to non-Watson-Crick base pair formation with the third base in the codon-anticodon interaction. These unconventional base pairs differ in structure from A·U and G·C, and the term "wobble hypothesis" suggests that there is some flexibility or "wobbling" allowed at this position in the ribosome.

Mismatches during DNA replication can arise due to the ability of DNA bases to form Watson-Crick base pairs as well as wobble base pairs. These mismatches can cause cancer and hereditary diseases if they are not fixed by DNA repair enzymes.

Did you get it?

1. **Focus on transcription: DNA to RNA**

1.4. GENE AND MECHANISM OF GENE ACTION

Learning Objectives

- Focus on classical concept of gene
- Explain the gene types;

Mendel's hybridization experiment has led to our understanding of genes as hereditary units linked to character inheritance that are passed down from one generation to the next via germ cells.

The term gene was first used by Johansen in 1909. After him, several workers experimentally proved that genes are the fractions or parts of DNA molecules are regarded as the genetic material. Suton and Boveri independently suggested that chromosome is the container of hereditary units.

T.M Morgan proposed the gene theory which states that:

- Chromosomes are bearers of hereditary units and each chromosome carries hundreds or thousands of genes.
- The genes are arranged on the chromosomes in a linear order and on the special regions or locus.

1.4.1. Classical Concept of Gene

The classical concept of genes was introduced by Sutton (1902) and was elaborated by Morgan (1913). Calvin Blackman Bridges (1923), Muller (1927), etc., is outlined as follows:

- Genes are discrete particles inherited in a mendelian fashion that occupy a definite locus in the chromosome and are responsible for the expression of a specific phenotypic character.
- The number of genes in each organism is more than the number of chromosomes; hence several genes are located on each chromosome.
- The genes are arranged in a single linear order like beads on a string.

- Each gene occupies a specific position called locus.
- If the position of the gene changes, the character changes.
- Genes can be transmitted from parent to off springs.
- Genes may exist in several alternate formed called alleles.
- Genes are capable of being combined together or can be replicated once during a cell division.
- Genes may suddenly changes in position and composition called mutation.
- Genes are capable of self-duplication producing their own exact copies.
  - **Modern Concept of Gene:** S. Benzer coined different terms for different natures of a gene and genetic material in relation to the chromosome on the basis of genetic phenomena that are involved.
  - **Genes as Unit of Transmission or Cistron:** The part of DNA specifying a single polypeptide chain is termed as cistron. A cistron can have 100 nucleotide pairs in length to 30,000 nucleotide pairs. It transmits characters from one generation to another as a unit of transmission.
  - **Genes as Unit of Recombination or Recon:** The smallest segment of DNA capable of being separated and exchanged with other chromosomes is called recon. A recon consists of not more than two pairs of nucleotides.
  - **Gene as Unit of Mutation or Muton:** Muton is the smallest unit of genetic material which when changed or mutated produces a phenotypic trait. Thus, the muton is delimited to a single nucleotide.

Keyword

Enzymes are proteins that help speed up metabolism, or the chemical reactions in our bodies.

1.4.1.1. Gene Action: (One Gene One Enzyme Concept)

Gene action is the expression of a genetic trait as a result of a gene's influence. Genes are typically linked to the synthesis of enzymes, which the body produces by autocatalyzing the synthesis of chemicals present in its cells. One gene typically influences one enzyme. The way that pigments develop is one way that genes express themselves in different ways. Colors, hormones, size and form, production of proteins, antigen and antibody production decisive effects on human diseases like albinism, tyrosinosis, etc. This idea was given by Beadle and Tatum.

1.4.1.2. Genes in Protein Synthesis and Genetic Code

Proteins of a particular type are produced by polypeptides with different types of amino acids found in a joint sequence. Its cell's chromosomes contain genes that directly regulate the synthesis of proteins. Codons, which are triplet bases, are used to transcribe specific m-RNA from chromosomes at the nucleus. There are 64 codons assigned for 20 distinct amino acid sequences. The aminoacid methionine in the polypeptide sequence is assigned by the codon "AUG" on mRNA. Jacob and Monod proposed the "Operon model" to explain the mechanism of gene action in protein synthesis is regulated by three specific genes located on chromosomes.

- **Structural Genes:** It regulates the production of a specific m-RNA and determines the kind of protein to be synthesized.
- **Operator Genes:** These genes act as switches to turn on or turn off the activities of structural genes, regulating the elongation and termination of a polypeptide chain.
- **Regulator Genes:** These genes produce certain proteinaceous substances called repressors, which prevent the operator genes from their action.

1.4.2. Gene Types

On the basis of their behavior, the genes may be categorized into the following types:

- **Basic Genes:** These are the fundamental genes that bring about the expression of a particular character.
- **Lethal Genes:** These bring about the death of their possessor.
- **Multiple Genes:** When two or more pairs of independent genes act together to produce a single phenotypic trait.
- **Cumulative Genes:** Some genes have additive effects on the action of other genes. These are called cumulative genes.
- **Pleiotropic Genes:** The genes that produce changes in more than one character are called pleiotropic genes.
- **Modifier Gene:** The gene which cannot produce a character by itself but interacts with others to produce a modified effect is called a modifier gene.
- **Inhibitory Gene:** The gene that suppresses or inhibits the expression of another gene is called the inhibitory gene.

Did you get it?

1. **What do you mean by gene?**
2. **Discuss about gene action: (one gene one enzyme concept).**

CASE STUDY

UNRAVELING GENETIC MYSTERIES: A CASE STUDY ON HUNTINGTON'S DISEASE

Introduction: Huntington's disease (HD) is a devastating neurodegenerative disorder caused by a genetic mutation affecting the huntingtin (HTT) gene. This case study delves into the molecular genetics of HD, exploring the underlying genetic mechanisms, clinical manifestations, and therapeutic approaches.

Case Presentation: Sarah, a 40-year-old woman, presents with involuntary movements, cognitive decline, and behavioral changes, along with a family history of similar symptoms. After thorough evaluation, Sarah is diagnosed with Huntington's disease based on clinical features and genetic testing revealing an expanded CAG repeat in the HTT gene.

Discussion:

- **Genetic Basis of Huntington's Disease:** HD is an autosomal dominant disorder caused by an expanded trinucleotide repeat sequence of cytosine-adenine-guanine (CAG) in the HTT gene located on chromosome 4. Normally, the HTT gene contains fewer than 35 CAG repeats. However, in individuals with HD, the CAG repeat expansion exceeds 36 repeats, leading to the production of mutant huntingtin protein (mHTT), which is toxic to neurons.
- **Pathogenesis of Huntington's Disease:** The exact pathogenic mechanisms underlying HD are complex and multifaceted. However, several key processes contribute to neuronal dysfunction and degeneration:
  - – **Aggregation of mHTT:** Expanded CAG repeats in the HTT gene result in the production of mHTT protein, which forms aggregates within neurons, disrupting cellular function and signaling pathways.
  - **Mitochondrial Dysfunction:** mHTT interferes with mitochondrial function, leading to oxidative stress, energy depletion, and impaired neuronal metabolism.

  Dysregulation of Gene Expression: mHTT alters the expression of various genes involved in synaptic plasticity, neurotransmitter signaling, and neuronal survival, contributing to neurodegeneration.
- **Clinical Manifestations:** Huntington's disease is characterized by progressive motor, cognitive, and psychiatric symptoms, including:
  - **Chorea:** Involuntary, jerky movements affecting the limbs, face, and trunk.
  - **Cognitive Decline:** Impairments in memory, attention, executive function, and visuospatial abilities.

  - **Psychiatric Symptoms:** Depression, anxiety, irritability, apathy, and psychosis.
  - **Functional Decline:** Loss of independence in activities of daily living, leading to significant disability and dependence on caregivers.
- **Diagnosis and Management:** Diagnosis of HD is based on clinical evaluation, family history, and genetic testing to detect the presence of CAG repeat expansions in the HTT gene. While there is currently no cure for HD, management strategies focus on symptomatic relief, supportive care, and multidisciplinary interventions:
  - **Pharmacological Treatments:** Medications targeting motor symptoms, psychiatric symptoms, and behavioral disturbances.
  - **Allied Health Interventions:** Physical therapy, occupational therapy, speech therapy, and nutritional support to optimize functional abilities and quality of life.
  - **Psychosocial Support:** Counselling, support groups, and caregiver education to address the emotional, social, and practical challenges faced by individuals and families affected by HD.

Conclusion: Huntington's disease exemplifies the intricate interplay between genetic factors, molecular pathways, and clinical phenotypes in neurodegenerative disorders. Understanding the molecular genetics of HD is essential for accurate diagnosis, prognostication, and the development of targeted therapies aimed at slowing disease progression and improving outcomes for affected individuals and their families. Ongoing research efforts continue to unravel the complexities of HD pathogenesis, offering hope for future breakthroughs in treatment and prevention strategies.

SUMMARY

- Molecular genetics is also essential to many other fields of biotechnology and biological research. It offers opportunities for the development of diagnostic instruments, treatments, and preventive measures by shedding light on the molecular causes of diseases.
- Nucleic acids, such as deoxyribonucleic acid (DNA) and ribonucleic acid (RNA), are necessary for the storage and transfer of genetic information in living things.
- Deoxyribonucleic acid, or DNA, is the genetic material found in humans and nearly all other living things. The DNA of an individual is found in almost all of their cells.
- The ability of DNA to replicate, or create duplicates of itself, is its crucial characteristic. The base sequence can be duplicated using a pattern found on each strand of DNA in the double helix. This is crucial during cell division because every newly formed cell requires an identical replica of the DNA found in the parent cell.
- The organic materials called nucleic acids found in all organisms are in the form of DNA or RNA. These nucleic acids are created through the combination of sugar molecules, phosphate groups, and nitrogenous bases joined by various bonds in a sequence.
- The process of copying DNA to mRNA, which contains the instructions required for protein synthesis, is known as transcription. There are two main stages to transcription. RNA polymerase enzymes are first involved in the formation of pre-messenger RNA.

MULTIPLE CHOICE QUESTIONS

1. **The process by which DNA is copied to produce an RNA molecule is called:**
   a. Translation
   b. Replication
   c. Transcription
   d. Reverse transcription
2. **Which type of RNA carries the genetic information from the DNA to the ribosomes for protein synthesis?**
   a. Messenger RNA (mRNA)
   b. Transfer RNA (tRNA)
   c. Ribosomal RNA (rRNA)
   d. Small nuclear RNA (snRNA)

3. **What is the name of the process by which RNA molecules are converted into proteins?**
   a. Replication
   b. Translation
   c. Transcription
   d. Reverse transcription
4. **In eukaryotic cells, where does transcription occur?**
   a. Cytoplasm
   b. Ribosomes
   c. Nucleus
   d. Endoplasmic reticulum
5. **What is the central dogma of molecular biology?**
   a. DNA is translated into RNA, which is translated into proteins.
   b. DNA is replicated during cell division.
   c. RNA is transcribed into DNA, which is translated into proteins.
   d. Proteins are translated into DNA, which is transcribed into RNA.

Answers to Multiple Questions

1. (c); 2. (a); 3. (b); 4. (c); 5. (a)

REVIEW QUESTIONS

1. Define a gene and explain its role in molecular genetics.
2. What is DNA?
3. What do you understand by RNA splicing?
4. Discuss about properties of genetic code.
5. Focus on Genes in protein synthesis and genetic code

REFERENCES

1. Abul-Husn, N. S., & Kenny, E. E. (2019). Personalized medicine and the power of electronic health records. *Cell*, *177*(1), 58–69.
2. Chen, R., & Snyder, M. (2013). Promise of personalized omics to precision medicine. *Wiley Interdisciplinary Reviews: Systems Biology and Medicine*, 5(2), 73–82.
3. Gibson, G. (2012). Rare and common variants: Twenty arguments. *Nature Reviews Genetics*, *13*(2), 135–145.
4. Hindorff, L. A., Sethupathy, P., Junkins, H. A., Ramos, E. M., Mehta, J. P., Collins, F. S., & Manolio, T. A. (2009). Potential etiologic and functional implications of

genome-wide association loci for human diseases and traits. *Proceedings of the National Academy of Sciences*, *106*(23), 9362–9367.

5. Karczewski, K. J., & Snyder, M. P. (2018). Integrative omics for health and disease. *Nature Reviews Genetics*, *19*(9), 499–512.
6. Manolio, T. A., Collins, F. S., Cox, N. J., Goldstein, D. B., Hindorff, L. A., Hunter, D. J., McCarthy, M. I., et al. (2009). Finding the missing heritability of complex diseases. *Nature*, *461*(7265), 747–753.
7. McCarthy, M. I., Abecasis, G. R., Cardon, L. R., Goldstein, D. B., Little, J., Ioannidis, J. P., & Hirschhorn, J. N. (2008). Genome-wide association studies for complex traits: Consensus, uncertainty, and challenges. *Nature Reviews Genetics*, *9*(5), 356–369.
8. Stegmann, U. E. (2016). Genetic coding reconsidered: An analysis of actual usage. *The British Journal for the Philosophy of Science*, *67*(3), 707–730. https://doi.org/10.1093/bjps/axv007.
9. Strauss, B. S. (2019). Martynas Yčas: The "archivist" of the RNA Tie club. *Genetics*, *211*(3), 789–795. https://doi.org/10.1534/genetics.118.301754.
10. Turanov, A. A., Lobanov, A. V., Fomenko, D. E., Morrison, H. G., Sogin, M. L., Klobutcher, L. A., et al. (2009). Genetic code supports targeted insertion of two amino acids by one codon. *Science*, *323*(5911), 259–261. https://doi.org/10.1126/science.1164748.
11. Vassy, J. L., McLaughlin, H. M., MacRae, C. A., Seidman, C. E., Lautenbach, D., Krier, J. B., Lane, W. J., et al. (2015). A one-page summary report of genome sequencing for the healthy adult. *Public Health Genomics*, *18*(2), 123–129.
12. Visscher, P. M., Wray, N. R., Zhang, Q., Sklar, P., McCarthy, M. I., Brown, M. A., & Yang, J. (2017). 10 years of GWAS discovery: Biology, function, and translation. *The American Journal of Human Genetics*, *101*(1), 5–22.
13. Wray, N. R., Yang, J., Hayes, B. J., Price, A. L., Goddard, M. E., & Visscher, P. M. (2013). Pitfalls of predicting complex traits from SNPs. *Nature Reviews Genetics*, *14*(7), 507–515.

CHAPTER 2

Chromosomal Basis of Inheritance

LEARNING OBJECTIVES

After studying this chapter, you will be able to:

- Give an overview of the chromosomal basis of inheritance;
- Discuss the chromosome structure and organization;
- Chromosomal mutations: duplication, deletion, inversion, and translocation
- Discuss about mosaicism and chimerism

INTRODUCTORY EXAMPLE

Fundamental Concept in Chromosomal Basis of Inheritance:

The chromosomal basis of inheritance is a fundamental concept in genetics that elucidates how genes are transmitted from one generation to the next via chromosomes. To illustrate this concept, let's consider a hypothetical scenario involving a family of cats with distinct fur colors.

Imagine a family of cats with a mother cat that has black fur and a father cat with white fur. These cats possess genes that determine their fur color, with the allele for black fur (B) being dominant over the allele for white fur (b). Each parent contributes one allele to their offspring, determining the fur color of the kittens.

At the cellular level, these genes for fur color are located on chromosomes within the cat's cells. Cats, like humans and many other organisms, have pairs of chromosomes, with one set inherited from each parent. In our scenario, let's simplify and consider a single pair of chromosomes, one inherited from the mother and one from the father.

The chromosome inherited from the mother may carry the allele for black fur (B), while the chromosome inherited from the father may carry the allele for white fur (b). During the formation of reproductive cells (gametes) through meiosis, these chromosomes segregate randomly, ensuring genetic diversity in the offspring.

Now, let's follow the inheritance of fur color in the kittens. Each kitten inherits one chromosome from each parent, resulting in combinations of alleles that determine their fur color. For example:

- If a kitten inherits the B allele from the mother and the b allele from the father (Bb), it will have black fur. This is because the dominant allele for black fur (B) masks the expression of the recessive allele for white fur (b).
- If a kitten inherits the B allele from both parents (BB), it will also have black fur.
- If a kitten inherits the b allele from both parents (bb), it will have white fur, as the recessive allele for white fur (b) is expressed in the absence of the dominant allele for black fur (B).

By observing the fur colors of the kittens, we can see how the chromosomal basis of inheritance governs the transmission of traits from parents to offspring. The chromosomes serve as vehicles for carrying genes, and the inheritance patterns observed follow the principles of Mendelian genetics, with alleles segregating and assorting independently during gamete formation. This example highlights how the chromosomal basis of inheritance underpins the diversity and inheritance of traits in populations of organisms.

UNIT INTRODUCTION

In genetics, the chromosomal basis of inheritance is a basic concept that describes how genes are passed down from parents to their children. This idea clarifies the relationship between Mendelian inheritance patterns and chromosomal behavior during cell division. It was initially put forth by Thomas Hunt Morgan and his associates through their ground breaking work with fruit flies in the early 20th Century. The chromosomal basis of inheritance also explains the phenomenon of genetic linkage, whereby genes located on the same chromosome tend to be inherited together due to their physical proximity. Meiotic recombination, also known as crossing over, can reorganize genetic material between homologous chromosomes, resulting in the formation of novel allele combinations. The chromosomal basis of inheritance, in its whole, emphasizes the significance of chromosomes in the organization and transmission of genetic information and offers a framework for comprehending the inheritance of genetic traits from one generation to another. There are significant ramifications for genetics, developmental biology, and evolutionary research from this idea.

Chromosomes are the home of genes. Genes, or their alleles, are found on each pair of chromosomes. Chromosomes are paired. Every chromosome ends up in a separate cell during the cell's half-divide. When sperm egg cells are formed during meiosis, this is observed. Additionally, the genes are divided in two. The scientific term for these is alleles. Only one homologous chromosome, not both, is present in meiotic products. Half of the chromosomes in haploid cells are produced during the process of meiosis, which is a sequence of cell divisions. Upon fertilization, the sperm and egg combine to form new pairs, but their genetic makeup is changed. Therefore, the mother contributes one allele while the father contributes the other. It is also through this mechanism that a faulty gene that results in a genetic disorder is passed down to progeny.

2.1. OVERVIEW OF CHROMOSOMAL BASIS OF INHERITANCE

Learning Objectives

- What are chromosomes?
- Explain the chromosomal theory of inheritance.

The relationship that exists between the chromosomal basis of inheritance and the chromosomes' role in the transfer of genetic information in living things is known as the inheritance patterns. The theory was proposed by Thomas Hunt Morgan and his colleagues through their experiments with fruit flies in the early 20th century.

DNA and proteins combine to form the thread-like structures called chromosomes, which are located in the nucleus of eukaryotic cells. Chromosomes contain genes, which are the basic building blocks of heredity and are in charge of encoding particular traits. The chromosomal behavior during meiosis, the cell division process that yields gametes (eggs and sperm), accounts for the inheritance patterns seen in progeny.

Meiosis produces genetic variation when homologous chromosomes randomly segregate into gametes. Furthering the increase in genetic diversity is the process of crossing over, in which homologous chromosomal segments exchange genetic material. In conjunction with the chromosomal theory of inheritance, Mendel's laws of segregation and independent assortment offer a thorough explanation of how traits are inherited and handed down through the generations (Figure 2.1).

Genetic diseases such as Turner syndrome and Down syndrome can be caused by abnormalities in the number or structure of chromosomes. The roles that particular genes on chromosomes play in a variety of physiological processes and diseases have also been clarified by advances in molecular genetics, which have improved our knowledge of human genetics and offered new perspectives on possible treatments for genetic disorders. In general, the chromosomal basis of inheritance serves as the cornerstone of contemporary genetics, directing studies in the fields of heredity and evolutionary biology.

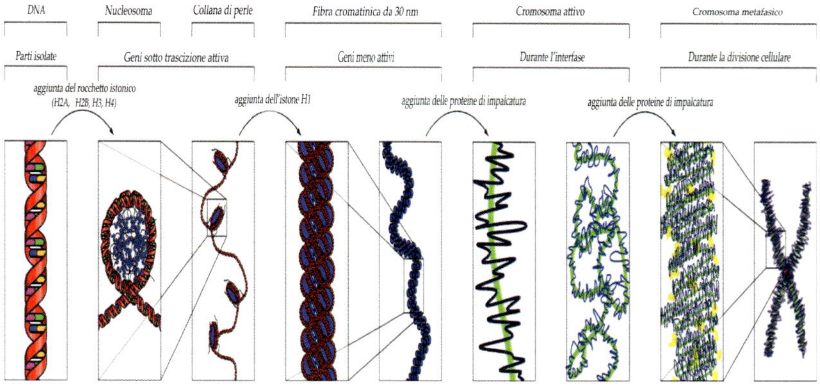

Figure 2.1. *The genetic information is primarily carried out by chromosomes.*

Source: By Original uploader was Richard Wheeler at en.wikipedia – Transferred from en.wikipedia to Commons by sevela.p., CC BY-SA 3.0, https://commons.wikimedia.org/w/index.php?curid=4017531.

2.1.1. What Are Chromosomes?

The punchline, "Genes lie on chromosomes," is probably familiar to you. But where do genes end up in a cell? The second punchline—that which introduced the modern genetic era—that genes are DNA segments that specify proteins may even be familiar to you.

Gene-shaped genetic information is primarily carried out by chromosomes, a thread-like structure made of nucleic acids and proteins that are found inside the nucleus of living cells. The Swiss botanist Carl Wilhelm von Nageli discovered chromosomes and was the first person to research the cell divisions.

Keyword

Human Genetics – It is the study of inheritance as it occurs in human beings.

2.1.2. The Chromosomal Theory of Inheritance

It was in the early 1900s that Boveri and Sutton presented the chromosomal theory of inheritance. It is the basic hypothesis of genetics. These theories state that chromosomes contain genes, which are the basic building blocks of heredity.

Mendelian genetics was discovered long before the chromosomal theory of inheritance. The society did not accept such radical shifts in scientific beliefs during Mendel's experiments. Since this contradicted their theory that evolution is the result of constant change, they found it hard to accept the existence of discrete factors like genes that would segregate without mixing. In addition, the lackluster communication infrastructure of the era prevented the general public from receiving the information. Mendel's use of mathematics to demonstrate biological laws was also inappropriate.

As time passed, scientists Vries, Correns, and Tschermak discovered that chromosomes existed inside the nucleus. Sutton and Boveri observed the behavior of the chromosomes when the cells were divided. With the advancements in the microscope, this task became easier. Hence, they proved Mendel's laws with the help of chromosomal movement. They showed the segregation of the chromosomes during the Anaphase of cell division. The idea of chromosomal segregation combined with the Mendelian principles gave rise to the chromosomal theory of Inheritance. The work was further carried forward and proved by T.H. Morgan, who used *Drosophila melanogaster* to show how sexual reproduction gave rise to variations.

It is the fundamental genetic theory that recognizes chromosomes as the bearers of genetic material. Theodor Boveri and Walter Sutton are the two scientists who have been credited with developing the Inheritance Chromosomal Theory. In the early 1900s, Boveri and Sutton gave the chromosomal theory of inheritance. It is the fundamental genetic theory. The chromosomes contain genes, which are units of heredity, according to this theory. Mendelian genetics was developed long before the inheritance chromosomal theory. At the time of Mendel's experiments, society was not open to such radical shifts in scientific beliefs. Not only did this contradict their theory of ongoing evolutionary changes, but they also did not think that distinct variables like genes could separate without combining.

In addition, the means of communication at that time were weak, as a result of which the information could not be transmitted to the masses. Mendel's scientific approach to proving biological rules has always been inappropriate. Scientists Vries, Correns, and Tschermak found chromosomes that remained within the nucleus as time passed. Sutton and Boveri discovered when the cells were divided they observed the behavior of the chromosomes. This task became simpler with the advances in the microscope. Thus, they used chromosomal motion to prove the laws of Mendel. Chromosome segregation during the anaphase of cell division was demonstrated. The Mendelian principles and the concept of chromosomal segregation have led to the development of Inheritance's Chromosomal Theory.

Did you get it?

1. **Discuss about mendelian genetics**

2.2. CHROMOSOME STRUCTURE AND ORGANIZATION

Learning Objectives

- Definition of chromosome;
- Discuss about functions of chromosomes.
- Explain the chromosome banding

The structures in cells that house genetic material are called chromosomes. Chromosomes in eukaryotic cells are made of DNA that is tightly wound around proteins known as histones to form a complex called chromatin. Chromosomes are primarily responsible for storing, safeguarding, and accurately transmitting genetic information during cell division.

The fundamental building block of chromatin is the nucleosome, which is made up of a DNA strand encircling eight histone protein cores. Chromatin is a string-like fiber made up of nucleosomes joined by additional histone proteins and linker DNA. During cell division, chromatin can continue to condense to create thicker fibers and ultimately compact into visible structures known as chromosomes.

Depending on the stage of the cell cycle, the chromosome arrangement can change. Chromosomes can express genes and replicate DNA when they are in the less compacted state of euchromatin, which occurs during interphase. Under a microscope, chromosomes become highly compacted and visible as cells get ready to divide. This process of condensation occurs repeatedly in cells.

Within the nucleus, chromosome arrangement is non-random. They occupy particular regions and engage with different nuclear structures such as the nucleolus, nuclear envelope, and other chromosomes. The arrangement of chromosomes in the nucleus can have an impact on the processes of DNA replication, gene expression, and repair (Figure 2.2).

Remember

Chromosome segregation into daughter cells during mitosis and meiosis is facilitated by this highly condensed state.

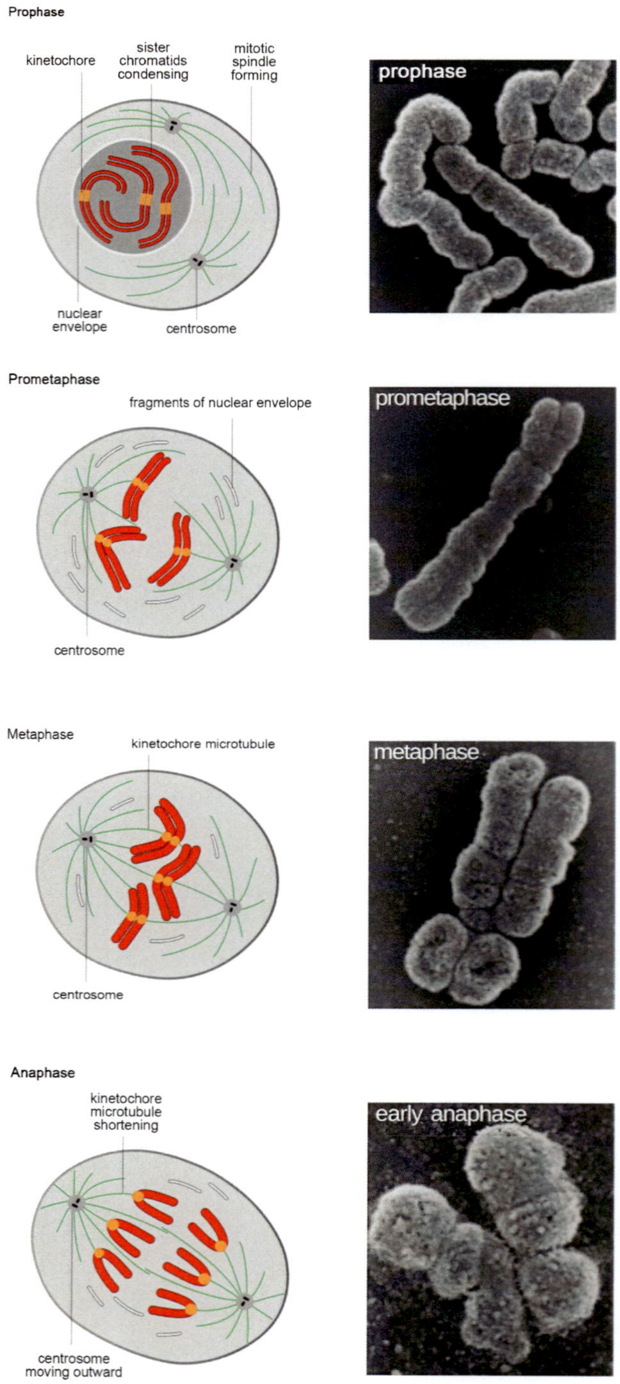

Figure 2.2. *Chromosomes: Structure, morphology, composition, and organization.*

Source: By David O Morgan – The Cell Cycle. Principles of Control, Attribution, https://commons.wikimedia.org/w/index.php?curid=89955214.

Telomeres and centromeres are unique regions found on chromosomes, in addition to their structural functions. For spindle fibers to attach during cell division and guarantee proper chromosomal segregation, centromeres are necessary. Protecting chromosomes from fusion and degradation are telomeres, which are found at their ends. Moreover, they are involved in the control of aging and cellular senescence.

Overall, the stability of the genome, control over gene expression, and precise transfer of genetic information from one generation to the next depend on the arrangement and structure of chromosomes. Genomic instability and several diseases, such as cancer and developmental disorders, can be caused by dysregulation of chromosomal structure and organization.

Keyword

Genome – It is the complete set of genetic information in an organism.

2.2.1. Definition of Chromosome

A chromosome is a coiled protein structure that resembles a thread. All cells have chromosomes in their nuclei, which house DNA, the basic genetic material that is passed down from one generation to the next.

2.2.1.1. Structure of Chromosomes

A chromosome has generally 8 parts; Centromere or primary constriction or kinetochore, chromatids, chromatin, secondary constriction, telomere, chromomere, chromonema, and matrix:

1. **Centromere or Kinetochore:** It is the primary constriction at the center to which the chromatids or spindle fibers are attached. Its function is to enable the movement of the chromosome during the anaphase stage of cell division.
2. **Chromatid:** During cell division, a chromosome is divided into 2 identical half strands joined by a centromere. A chromatid is each half of the chromosome joined. Each chromatid contains DNA and separates at Anaphase to form a separate chromosome. Both chromatids are attached to each other by the centromere.
3. **Chromatin:** It is a complex of DNA and proteins that forms chromosomes within the nucleus of eukaryotic cells. Nuclear DNA is highly condensed and wrapped around nuclear proteins in order to fit inside the nucleus. In other words, it is not present as free linear strands. The chromatin consists of DNA, RNA, and protein.
4. **Secondary Constriction:** It is generally present for the nucleolar organization.

5. **Telomere:** It is the terminal region of each side of the chromosome.
6. **Chromonema:** It is a threadlike coiled filamentous structure along which chromomeres are arranged. Chromonema controls the size of the chromosome and it acts as a site of gene bearing.
7. **Chromomeres:** These are the bead-like structures present on threads or chromonema. These are arranged in a row along the length of chromonema. The number of chromosomes is constant and it is responsible for carrying the genes during cell division to the next generation.
8. **Matrix:** Pellicle is the membrane surrounding each of the chromosomes. Matrix is the jelly-like substance present inside the pellicle. It is formed of non-genetic materials.

2.2.2. Functions of Chromosomes

For the first time, Sutton and Bover suggested the role of chromosomes in heredity in 1902:

1. Condensation of nuclear material? Each chromosome stretched has the length of 2500 human cells. Histones and other accessory proteins have the function of condensing the DNA in a way that the cell can survive.
2. Histones and other proteins cover the Chromosomes. These proteins protect it from chemical (e.g., enzymes) and physical forces. Thus, chromosomes also perform the function of protecting the genetic material (DNA) from damage during the process of cell division.
3. During cell division, spindle fibers attached to the centromeres contract and perform an important function. The contraction of centromeres of chromosomes ensures the precise distribution of DNA (genetic material) to the daughter nuclei.
4. Chromosomes contain histone and non-histone proteins, these proteins regulate gene action. Cellular molecules that regulate genes work by activating or deactivating these proteins. This activation and deactivation expand or contract the chromosome.

Examples of Types of Chromosomes

1. **Metacentric Chromosomes:** These have the centromere present exactly in the center. Both sections are metacentric chromosomes and are therefore of equal length. Example: Human chromosomes 1 and 3 are metacentric.
2. **Submetacentric Chromosomes:** In this, the centromere is not present exactly at the center. The centromere is slightly offset from the center. Both the sections are therefore not of equal length or are asymmetrical. Example: Human chromosomes 4 to 12 are submetacentric.
3. **Acrocentric Chromosomes:** These have a centromere that is highly offset from the center. Therefore, one of the strands is very long and one very short. Example: Human chromosomes 13,15, 21, and 22 are acrocentric.
4. **Telocentric Chromosomes:** In this, the centromere is present at the

very end of the chromosome. Telocentric chromosomes are present in species such as mice. Humans do not possess telocentric chromosomes.

Did you know?

In a series of experiments beginning in the mid-1880s, Theodor Boveri gave definitive contributions to elucidating that chromosomes are the vectors of heredity, with two notions that became known as 'chromosome continuity' and 'chromosome individuality.'

2.2.3. Chromosome Banding

There are several types of chromosome banding. Here, we will list a few of the most common types:

1. **Q-Banding:** The alkylating agent quinacrine mustard was the first substance to band chromosomes observed through a fluorescence microscope. Quinacrine mustard has since been used in place of quinacrine dihydrochloride. Q bands are the bright and dull fluorescence bands that alternate. Whereas the dull bands are rich in guanine and cytosine, the bright bands are primarily made of DNA rich in adenine and thymine.

   Q bands are especially useful for distinguishing the human Y chromosome and various chromosome polymorphisms involving satellites and centromeres of specific chromosomes.

2. **G-Banding:** The stain that is now most frequently used in human cytogenetic analysis is called Giemsa. G-banding, in contrast to Q-banding, typically necessitates pre-treating chromosomes with an enzyme that breaks down proteins or salt. GTG banding is the term for the procedure that occurs when chromosomes are pre-treated with the proteolytic enzyme trypsin. Giemsa preferentially stains areas high in thymine and adenine. Therefore, G bands correspond closely to Q bands.

   Metaphase chromosomes can show 400–600 bands using standard G band staining techniques. 24 human chromosomes have up to 2,000 distinct bands that have been cataloged using high-resolution G-banding techniques.

3. **R-Banding:** Giemsa staining and hot phosphate buffer incubation are steps in the reverse banding (R-banding) process for slides containing metaphase chromosomes. This leads to a banding pattern that is basically G bands in reverse. GC is abundant in R bands. Heat denatures the AT-rich regions only, leaving the GC-rich regions unaltered. GC-specific fluorescent proteins also result in a reverse chromosome banding pattern. Since chromosome ends typically stain light with G-banding, R-banding is useful for examining their structure.

4. **C-Banding:** It stains areas of heterochromatin, which is tightly packed and repetitive DNA. C-banding is specifically useful in humans to stain the centromeric chromosome regions and other regions containing constitutive heterochromatin –

secondary constrictions of human chromosomes 1, 9, 16, and the distal segment of the Y chromosome long arm.

5. **NOR-Banding:** It involves silver staining (silver nitrate solution) of the "nucleolar organizing region," which contains rRNA genes.
6. **T-Banding:** It is the process of staining telomeric regions of chromosomes with acridine orange or Giemsa following carefully regulated heat denaturation. Because T bands are more strictly telomeric and smaller than corresponding R bands, they appear to represent a subset of the R bands.

2.3. CHROMOSOMAL MUTATIONS: DUPLICATION, DELETION, INVERSION, AND TRANSLOCATION

Learning Objectives

- Focus on duplication
- Discuss about deletion

Changes to an organism's chromosomal makeup or number, known as chromosomal mutations, can have a profound impact on its phenotypic. Inversion, translocation, deletion, and duplication are the four most frequent forms of chromosomal mutations.

- **Duplication:** Duplication is the process of copying a chromosome segment more than once, producing an excess of genetic material. This can result in an imbalance in the dosage of certain genes, which can cause abnormal cellular processes due to elevated expression of those genes. Duplication events can result from chromosomal rearrangements or mistakes made during DNA replication.
- **Deletion:** A chromosome is lost when a segment is deleted. This leads to the loss of particular genetic material, which can cause gene regulation to be disrupted or gene function to be lost. The size of deletions can vary, encompassing large-scale deletions encompassing entire genes or chromosomal regions or small-scale deletions involving a few base pairs. Mutations can occur naturally or be brought on by triggers from the environment or pathogens.
- **Inversion:** When a chromosomal segment breaks and reattaches in the opposite orientation, it is known as inversion. The genetic sequence within the inverted region is reversible as a result. Gene inversions can cause disruptions to gene expression or protein structure by flipping the order or orientation of genes.

Errors in DNA repair processes can result in de novo emergence of reversals or inheritance of reversals.

- **Translocation:** A translocation is the transfer of a section of one non-homologous chromosome to another. Translocations can be classified into two main categories: non-reciprocal translocations, in which a segment is transferred from one chromosome to another without reciprocal exchange, and reciprocal translocations, in which two non-homologous chromosomes exchange segments. Genetic disorders or cancer may arise from translocations that cause gene disruption, gene fusion, or altered gene regulation.

These chromosomal mutations can disrupt gene function, change gene dosage, or affect gene regulation, all of which can have a substantial impact on an organism's phenotypic. Studying chromosomal mutations and their effects is crucial for understanding genetic disorders, evolutionary processes, and the creation of therapeutic interventions.

Did you get it?

1. What is inversion?
2. What do you understand by translocation?

2.4. MOSAICISM AND CHIMERISM

Learning Objectives

- Discuss about cytogenetic mosaics
- Explain the X chromosome mosaicism

Chimeras and mosaics are made up of multiple genetically distinct cell populations. Though it is occasionally disregarded or misapplied, the distinction between these two forms is fairly obvious. A single zygote gives rise to all the genetically distinct cell types in mosaics, while multiple zygotes give rise to chimeras. It is not unusual for mammals to be mosaics; in fact, about half of all mammals are mosaics. Conversely, unless you are a cattle farmer or an experimental embryologist, you are unlikely to encounter a chimera.

2.4.1. Cytogenetic Mosaics

In a human mosaic, some of the cells might be 46, XX and some 47, XXX. The fraction of cells having each genotype is quite variable, reflecting how early during embryogenesis the mosaicism originated. In most but not all cases, such mosaicism can be detected in cells from all tissues.

What does mosaicism mean clinically? A mosaic will show symptoms of disease if the percentage of cytogenetically abnormal cells in that individual is high enough. On the other hand, normal cells might be enough to stop disease or lessen its severity if the aberrant cells are proportionately smaller than cytogenetically normal cells.

2.4.2. X Chromosome Mosaicism

X chromosome inactivation is the process by which all but one X chromosome in mammals become functionally inactive early in embryogenesis. All normal females have roughly equal populations of two genetically distinct cell types because this inactivation happens randomly; as a result, they are a type of mosaic. The maternal X chromosome is inactive in half of their cells, and the paternal X chromosome is inactive in the other half. This has significant biological and medical ramifications, especially for disorders linked to X genes.

Cats offer a special chance to witness X chromosome inactivation and aid in illustrating how it impacts all females. The tortoiseshell cats have a coat that is a blend of orange and black hair. Similar in appearance, calico cats also have white patches, which are caused by a different gene. As can be seen below, tortoiseshell cats have a coat that is a blend of orange and black hair.

Keyword

A mosaic is a pattern or image made of small regular or irregular pieces of colored stone, glass or ceramic, held in place by plaster/mortar, and covering a surface.

With regard to the orange gene, female cats with two X chromosomes can have any one of three genotypes: OO (orange coat), oo (black coat), or Oo (tortoiseshell or calico). The pattern of X chromosome inactivation in the hair follicles is reflected in the tortoiseshell pattern of tiny patches of orange and black.

The non-orange allele-bearing X chromosome is active and the orange allele-bearing X chromosome is inactivated in black patches. Orange fur patches have the exact opposite appearance. The X chromosome inactivation is haphazard; while there are sizable areas of both black and orange (like five heads in a row when flipping a coin), the majority of the coat is a subtle blend of orange and black (heads, tails, tails, heads, tails).

What if you're not into cat coat colors? Perhaps you're not even a fan of cats? Yes, it is still important to understand that all female mammals have the same pattern of X chromosome inactivation, which is reflected in the black and orange fur found in their coats. Simply put, that pattern is typically invisible because X-linked genes do not, for example, encode the color of human skin. However, comprehending mosaicism and X chromosome inactivation in all species is crucial to comprehending the pathophysiology of X-linked genetic disorders.

2.4.3. Chimeras

A chimera is a fire-breathing monster that has the head of a lion, the body of a goat, and the tail of a serpent in mythology. According to medical science, a chimera is a person who has multiple genetically different cell populations that were formed from multiple zygotes.

It is by no means uncommon to find chimeric cattle. Anastomoses—or areas of joining—between the fetal circulatory systems develop early in gestation when a cow gives birth to twins. This is almost assured. Because of this, the two fetuses exchange blood. Stem cells from the twins are permanently "seeded" into each fetus, and fetal blood contains hematopoietic stem cells.

Reports exist of naturally-occurring chimerism in a range of species. While these instances are uncommon, they undoubtedly exist. The prevailing explanation for this phenomenon is the merging of two early embryos into a single entity. This theory gains support from the fact that chimeras can also be intentionally created in laboratory settings, proving to be a valuable research resource across various biomedical fields. The fundamental technique involves the combination of two extremely early-stage embryos, allowing their cells to mix, resulting in a conceptus comprised of cells originating from both original embryos. This technique has been widely applied to mice and has also been applied to ruminants.

Fundamental inquiries concerning developmental biology and pathology have been most effectively addressed by chimeras involving sheep and goats, as well as mice. Additionally, there's a chance that issues like the rescue of endangered species can be solved with this technique. For instance, a goat-sheep chimera could be created in which the placenta of a sheep "encases" a goat fetus. This allows a sheep to carry a goat to term, something that cannot happen if you just implant goat embryos into sheep because the sheep will reject the placenta and fetus on an immunological level. It might be feasible to expand this process so that recipient mothers of different species can carry embryos from critically endangered species.

Did you get it?

1. **What do you understand by chimeras?**

CASE STUDY

UNDERSTANDING THE CHROMOSOMAL BASIS OF INHERITANCE: A CASE STUDY ON DOWN SYNDROME

Introduction: Down syndrome, also known as trisomy 21, is a chromosomal disorder caused by the presence of an extra copy of chromosome 21. It is one of the most common genetic disorders, occurring in approximately 1 in every 700 births worldwide. This case study explores the chromosomal basis of Down syndrome, its genetic and phenotypic characteristics, and its impact on affected individuals and their families.

Case Presentation: Sarah, a 35-year-old woman, gives birth to a baby boy named Jack. Shortly after birth, Jack is diagnosed with Down syndrome based on clinical features such as characteristic facial appearance, developmental delays, and congenital heart defects. Sarah and her husband are shocked and confused by the diagnosis, and they seek further information about the condition.

Discussion:

- **Chromosomal Basis:** Down syndrome is primarily caused by the presence of an extra copy of chromosome 21, resulting from nondisjunction during meiosis. Instead of the normal pair of chromosome 21, individuals with Down syndrome have three copies of chromosome 21 (trisomy 21). This additional genetic material leads to the characteristic features and developmental challenges associated with the condition.
- **Genetic Mechanism:** Nondisjunction can occur during either maternal or paternal gamete formation, leading to the inheritance of an extra chromosome 21. In most cases (about 95%), the extra chromosome 21 arises from nondisjunction in the mother's egg cells. Advanced maternal age is a significant risk factor for nondisjunction, although the majority of babies with Down syndrome are born to younger mothers due to higher fertility rates in this age group.
- **Phenotypic Characteristics:** Individuals with Down syndrome typically exhibit a range of physical and developmental characteristics, including:
  - **Facial Features:** Upward slanting eyes, flattened facial profile, small ears, and protruding tongue.
  - **Developmental Delays:** Delayed milestones such as sitting, crawling, and walking, as well as intellectual disability.
  - **Medical Complications:** Congenital heart defects, gastrointestinal issues,

hearing and vision problems, and increased risk of leukemia and Alzheimer's disease later in life.

- **Diagnosis and Management:** Diagnosis of down syndrome can be made prenatally through screening tests such as ultrasound and maternal serum screening, followed by confirmatory diagnostic tests like amniocentesis or chorionic villus sampling. Postnatal diagnosis is based on clinical evaluation and chromosomal analysis. Management of Down syndrome involves multidisciplinary care, including early intervention programs, special education services, medical surveillance for associated health issues, and supportive therapies to address developmental and behavioral challenges.

Conclusion: Down syndrome exemplifies the chromosomal basis of inheritance, with the presence of an extra chromosome 21 leading to characteristic phenotypic features and developmental challenges. Understanding the genetic mechanisms underlying Down syndrome is crucial for accurate diagnosis, appropriate management, and support for affected individuals and their families. Advances in genetic testing and interventions continue to improve outcomes and quality of life for individuals with Down syndrome, highlighting the importance of ongoing research and advocacy in this field.

SUMMARY

- In genetics, the chromosomal basis of inheritance is a basic concept that describes how genes are passed down from parents to their children.
- Chromosomes are the home of genes. Genes, or their alleles, are found on each pair of chromosomes. Chromosomes are paired. Every chromosome ends up in a separate cell during the cell's half-divide.
- The primary function of chromosomes, a thread-like structure made of nucleic acids and proteins that is found inside the nucleus of living cells, is to carry genetic information in the form of genes.
- the means of communication at that time were weak, as a result of which the information could not be transmitted to the masses. Mendel's scientific approach to proving biological rules has always been inappropriate.
- Changes to an organism's chromosomal makeup or number, known as chromosomal mutations, can have a profound impact on its phenotypic.

MULTIPLE CHOICE QUESTIONS

1. **Which scientist is credited with discovering the chromosomal theory of inheritance?**
   a. Gregor Mendel
   b. Thomas Hunt Morgan
   c. Alfred Hershey
   d. James Watson
2. **In the chromosomal theory of inheritance, genes are located on:**
   a. Proteins
   b. Enzymes
   c. Chromosomes
   d. Lipids
3. **How many sets of chromosomes do humans typically have?**
   a. 23
   b. 46
   c. 22
   d. 44
4. **Which of the following chromosomes determines an individual's sex in humans?**
   a. Autosomes
   b. X chromosome
   c. Y chromosome
   d. Both X and Y chromosomes

5. In a typical human karyotype, males have:

a. Two X chromosomes

b. One X chromosome and one Y chromosome

c. Two Y chromosomes

d. One X chromosome

REVIEW QUESTIONS

1. What is the chromosomal theory of inheritance, and who proposed it?
2. Explain the difference between autosomes and sex chromosomes.
3. Describe the process of meiosis and its significance in the chromosomal basis of inheritance.
4. What is crossing over, and how does it contribute to genetic diversity?
5. How does nondisjunction during meiosis lead to chromosomal disorders such as Down syndrome, Turner syndrome, and Klinefelter syndrome?

Answers to Multiple Questions

1. (b); 2. (c); 3. (b); 4. (d); 5. (b)

REFERENCES

1. Berry, A., & Watson, J. (2003). *DNA: The Secret of Life*. New York: Alfred A. Knopf. ISBN 0-375-41546-7.
2. Calladine, C. R., Drew, H. R., Luisi, B. F., & Travers, A. A. (2003). *Understanding DNA: The Molecule & How it Works.* Amsterdam: Elsevier Academic Press. ISBN 0-12-155089-3.
3. Carina, D., & Clayton, J. (2003). *50 Years of DNA*. Basingstoke: Palgrave Macmillan. ISBN 1-4039-1479-6.
4. Chao, S., Elias, E., Benscher, D., et al. (2016). Genetic mapping of major-effect seed dormancy quantitative trait loci on chromosome 2B using recombinant substitution lines in tetraploid wheat. *Crop Science*, *56*(1).
5. Dahm, R. (2008). Discovering DNA: Friedrich Miescher and the early years of nucleic acid research. *Human Genetics*, *122*, 565–581.
6. Fay, D. (2006). Genetic mapping and manipulation: Chapter 1-Introduction and basics. *WormBook, 17*, 1–12.
7. Micklas, D. (2003). *DNA Science: A First Course*. Cold Spring Harbor Press. ISBN 978-0-87969-636-8.
8. Olby, R. C. (2009). *Francis Crick: A Biography*. Plainview, N.Y.: Cold Spring Harbor Laboratory Press. ISBN 978-0-87969-798-3.

9. Rich, A., & Zhang, S. (2003). Z-DNA: The long road to biological function. *Nature Reviews Genetics*, *4*, 566–572.
10. Ridley, M. (2006). *Francis Crick: Discoverer of the Genetic Code*. Ashland, OH: Eminent Lives, Atlas Books. ISBN 0-06-082333-X.
11. Rosenfeld, I. (2010). *DNA: A Graphic Guide to the Molecule that Shook the World*. Columbia University Press. ISBN 978-0-231-14271-7.
12. Schultz, M., & Cannon, Z. (2009). *The Stuff of Life: A Graphic Guide to Genetics and DNA*. Hill and Wang. ISBN 978-0-8090-8947-5.
13. Wolf, G. (2003). Friedrich Miescher: The man who discovered DNA. *Chemical Heritage*, *21*(10, 11), 37–41.

CHAPTER

3

Genetic Inheritance: Mendelian and Non-Mendelian Patterns

LEARNING OBJECTIVES

After studying this chapter, you will be able to:

- Discuss the mendelian inheritance
- Explain the non-mendelian inheritance
- Focus on multiple factor inheritance

INTRODUCTORY EXAMPLE

Genetics Is the Study of Genes and Tries to Explain What They Are and How They Work:

Genetics, the study of heredity and variation in living organisms, encompasses a vast array of fascinating phenomena that shape the diversity of life. An introductory example in genetics often begins with the study of simple traits, such as eye color in humans, which offers a glimpse into the complexity of genetic inheritance.

Consider the inheritance of eye color, a trait determined by multiple genes, including OCA2, HERC2, and others. These genes exist in different forms called alleles, which contribute to the range of eye colors observed in populations. For instance, the brown eye allele (B) is dominant over the blue (b) and green (g) alleles. This dominance establishes a hierarchy where individuals with at least one dominant allele express the dominant trait, while the recessive traits are expressed only when an individual inherits two copies of the recessive allele.

When exploring the inheritance of eye color, it becomes evident that genetic patterns are not always straightforward. For instance, if both parents have brown eyes, their offspring are likely to inherit brown eyes due to the dominant nature of the brown allele. Conversely, if both parents have blue eyes, their offspring will almost always have blue eyes, as they must inherit two copies of the recessive blue allele from their parents.

However, complications arise when considering mixed parental traits. If one parent has brown eyes (Bb) and the other has blue eyes (bb), their offspring could inherit either brown or blue eyes, depending on the alleles passed down from each parent during reproduction. This unpredictability underscores the intricate nature of genetic inheritance, where various combinations of alleles interact to produce diverse phenotypes.

Furthermore, genetic diversity extends beyond simple Mendelian genetics, as environmental factors and genetic mutations contribute to the complexity of traits observed in populations. Individuals may exhibit intermediate eye colors or unique patterns due to the interplay of genetic and environmental influences.

Studying introductory examples in genetics, such as the inheritance of eye color, not only provides insights into the mechanisms underlying heredity but also lays the groundwork for understanding more complex genetic phenomena. Concepts like allele inheritance, dominance, and the role of multiple genes in trait determination serve as building blocks for exploring the intricacies of genetic inheritance in diverse organisms.

UNIT INTRODUCTION

Genetic inheritance encompasses both Mendelian and non-Mendelian patterns, reflecting the complexity of how traits are passed from one generation to the next. Mendelian inheritance, named after Gregor Mendel, follows clear patterns based on dominant and recessive alleles. For instance, Mendel's pea plant experiments revealed that traits such as flower color are determined by a single gene with two alleles, where one can mask the effect of the other. This results in predictable ratios in offspring, exemplified by the 3:1 ratio seen in dominant-recessive traits.

Non-Mendelian inheritance, however, includes more complex patterns that do not fit Mendel's simple models. Examples include incomplete dominance, where heterozygous individuals display a blend of both parental traits, and codominance, where both alleles contribute equally and visibly to the phenotype. Epigenetics and environmental factors can also influence gene expression, further complicating inheritance patterns. These non-Mendelian mechanisms illustrate the intricate nature of genetic inheritance beyond Mendel's foundational principles.

3.1. MENDELIAN INHERITANCE

Learning Objectives

- General Principles of Inheritance

Mendelian inheritance is the term used to describe how two alleles of a single gene, one of which may be fully dominant over the other, are inherited. Whether a trait is regulated by genes on the sex chromosome or by genes on the autosome determines the inheritance pattern of Mendelian traits.

- Autosomal traits are controlled by genes on one of the 22 pairs of human autosomes. Autosomes are all the chromosomes except the X or Y chromosome, and they do not differ between males and females, so autosomal traits are inherited in the same way, regardless of the sex of the parent or offspring.
- Traits controlled by genes on the sex chromosomes are called sex-linked traits. Because of the small size of the Y chromosome, most sex-linked traits are controlled by genes on the X chromosome. These traits are called X-linked traits. Single-gene X-linked traits have a different pattern of inheritance than single-gene autosomal traits because males have just one X chromosome. In addition, males always inherit their X chromosome from their mother, and they pass on their X chromosome to all of their daughters, but none of their sons.

3.1.1. General Principles of Inheritance

Many analytical challenges arise from traits that are continuously changing. We will talk about how to analyze quantitative traits, the physiological processes that underlie them, subline divergence within inbred strains, heterosis, and inbreeding depression.

Genetic inheritance is governed by several key principles that describe how traits are passed from parents to offspring. The foundational concepts include:

- **Segregation**: This principle, derived from Gregor Mendel's work, states that each individual possesses two alleles for each trait, one inherited from each parent. During gamete formation (meiosis), these alleles separate so that each gamete carries only one allele. Upon fertilization, the offspring inherits one allele from each parent, restoring the pair.
- **Independent Assortment**: Mendel's second principle indicates that genes located on different chromosomes are inherited independently of one another. This

means the inheritance of an allele for one trait does not affect the inheritance of an allele for another trait, leading to genetic diversity.

- **Dominance**: In cases where two different alleles for a trait are present, one may be dominant over the other. The dominant allele masks the effect of the recessive allele, determining the organism's phenotype. For example, in pea plants, the allele for purple flowers is dominant over the allele for white flowers.
- **Uniformity**: When two homozygous parents with contrasting traits breed, all offspring in the first generation (F1) will exhibit the dominant trait, showing uniformity in phenotype.
- **Segregation of Alleles**: During meiosis, the two alleles for a trait segregate so that each gamete carries only one allele. This ensures that offspring inherit one allele from each parent, following Mendel's Law of Segregation.

These principles provide a framework for understanding the predictable patterns of inheritance, though modern genetics also reveals more complex interactions, such as gene linkage and environmental influences.

3.1.1.1. Studying Inheritance Patterns

There are two very useful tools for studying how traits are passed from one generation to the next. One tool is a pedigree, and the other is a Punnett square.

1. **Pedigree:** The chart below is called a pedigree chart (Figure 3.1). The inheritance pattern of a trait within a family is displayed in a pedigree. An autosomal or X-linked Mendelian trait, for instance, can be determined by looking at a pedigree. The genotypes of various family members can also be deduced using it. In an imaginary autosomal trait with a dominant allele, one can see symbols of a married couple at the top of the pedigree chart. Though the wife is not affected (unaffected female), the husband (affected male) possesses the trait. Three of the children's spouses are displayed in the next row of the pedigree along with the children themselves. A male affected by the condition and his unaffected wife are the parents of the first child on the left, for instance. The subsequent generation is represented in the pedigree's third row by the couple's grandchildren at the top. The afflicted female child on the left is the sibling of an unaffected male child in this generation.
2. **Punnett Square:** It is a chart that allows you to easily determine the expected ratios of possible genotypes in the offspring of two parents. If the gene is autosomal, and both parents are heterozygotes *(Aa)* for the gene. Half of the gametes produced by each parent will have the *A* allele, and half will have the *a* allele. That's because the two alleles are on homologous chromosomes, which always separate and go to separate gametes during meiosis. The alleles in the gametes from each parent are written down the side and across the top of the Punnett square. Filling in the cells of the Punnett square gives the possible genotypes of their children. It also shows the most likely ratios of the genotypes, which in this case is 25% *A,* 50% *Aa,* and 25% *a.* You can see a hypothetical example in Figure 3.2.

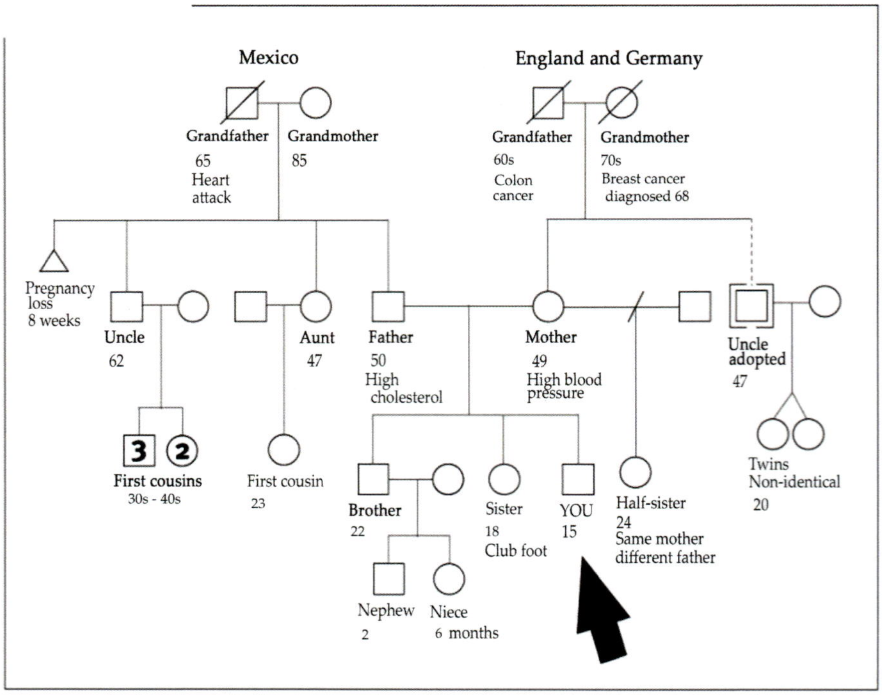

Figure 3.1. *A pedigree chart is similar to a family tree. It shows how a trait is passed from parents to offspring in a family.*

Source: https://courses.lumenlearning.com/wm-nmbiology1/chapter/reading-pedigrees-and-disease/.

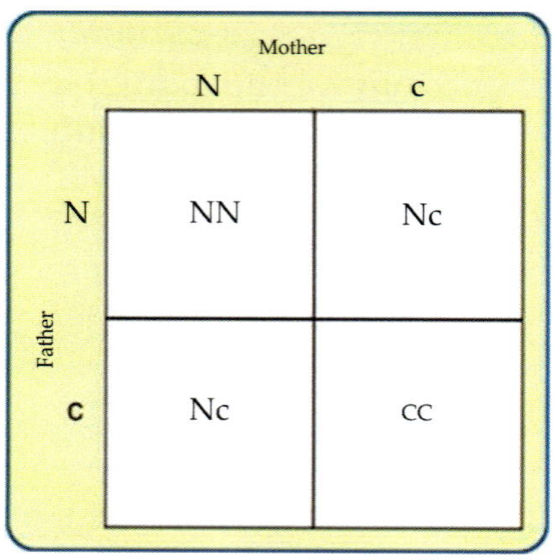

Figure 3.2. *A Punnett square shows the most likely proportions of offspring by genotype for a particular mating type.*

Source: https://courses.lumenlearning.com/wm-lifespandevelopment/chapter/heredity-and-chromosomes.

A Punnett square can also be used to show how the X and Y chromosomes are passed from parents to their children. This is illustrated in the Punnett square in Figure 3.3. It may help you understand the inheritance pattern of sex-linked traits.

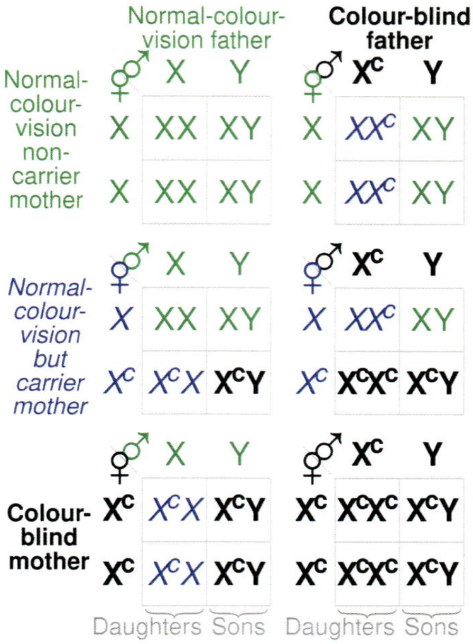

Figure 3.3. *Inheritance of sex chromosomes. Mothers pass only X chromosomes to their children. Fathers always pass their X chromosome to their daughters, and their Y chromosome to their sons. Can you explain why fathers always determine the sex of the offspring?*

Source: By Cmglee – Own work, CC BY-SA 4.0, https://commons.wikimedia.org/w/index.php?curid=106072201.

3.1.1.2. Autosomal Mendelian Traits in Humans

Single genes with two alleles control only a small percentage of human autosomal traits, but they provide a useful foundation for studying human heredity. Human albinism, in its most common forms, follows a Mendelian inheritance pattern, as was covered at the outset of this concept. Typically, a single person controls albinism autosomal gene with two alleles. The allele for normal pigmentation (let's call it *R*) is dominant to the allele for albinism (*r*). Individuals with either an *RR* or *Rr* genotype will not have albinism, because the *R* allele is dominant over the recessive *r* allele.

However, consider what happens if two individuals with the *Rr* genotype reproduce with each other. The outcome would be similar to the example shown in the Punnett square above for two hypothetical *Aa* individuals. Their possible offspring could be *R* (normal pigmentation), *Rr* (normal pigmentation), or *r* (albinism). This explains why a child with albinism (*r*), can have two parents that do not have albinism.

Keyword

Gene – It is the basic physical and functional unit of heredity.

- **Remember:** Both unaffected parents must be carriers of the recessive *r* allele, but they also have a dominant *R* allele that prevents them from having the condition themselves.

Other characteristics of humans that follow a Mendelian inheritance pattern include wet versus dry ear wax and Huntington's disease. You might be familiar with other characteristics of people that were once believed to be Mendelian, like having dimples, a widow's peak hairline, a hitchhiker's thumb, and the capacity to roll your tongue. It is now recognized by science that these are not truly Mendelian characteristics. The majority of human traits do not follow a straightforward Mendelian inheritance pattern; rather, they are influenced by several genes or otherwise contain more than two alleles.

3.1.1.3. X-Linked Mendelian Traits in Humans

For every X-linked trait, males have only one allele since they have one X chromosome. Males always express a recessive X-linked allele as a result. Every X-linked trait has two alleles in females since they have two X chromosomes. Therefore, for them to exhibit an X-linked recessive trait, they need to inherit two copies of the recessive allele. This elucidates the reason behind the lower frequency of X-linked recessive traits in females relative to males, as well as their distinct inheritance pattern when compared to autosomal traits. The color blindness of red and green is an illustration of a recessive X-linked trait. This condition prevents a person from telling the difference between the colors red and green. This characteristic, which is comparatively uncommon in females but fairly common in males, is encoded by multiple recessive genes on the X chromosome (Figure 3.4). Figure 3.5 shows a simple pedigree for this trait.

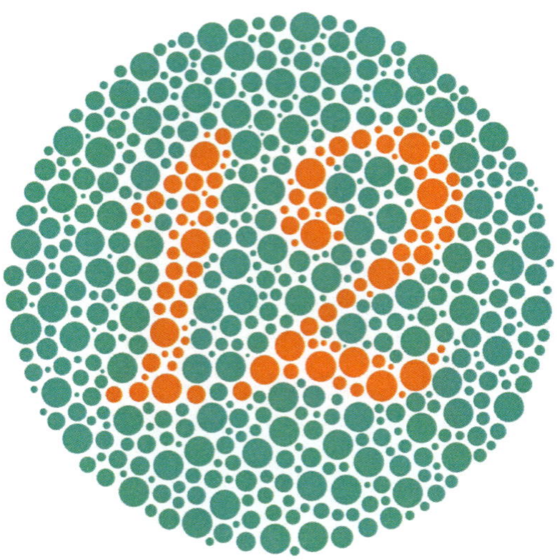

Figure 3.4. *This circle of colors containing the number 74 is part of the Ishihara color blindness test.*

Source: By Shinobu Ishihara – Unknown source, Public Domain, https://commons.wikimedia.org/w/index.php?curid=47050887.

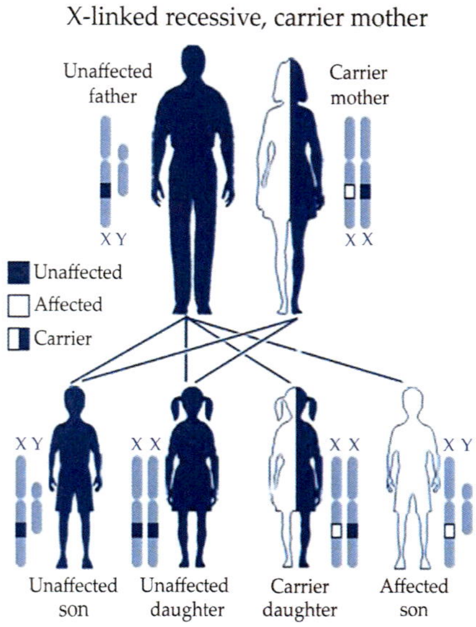

Figure 3.5. *Heredity implications of an X-linked recessive gene carried by the mother.*

Source: https://courses.lumenlearning.com/suny-dutchess-ap1/chapter/patterns-of-inheritance/: (credit: U.S. National Library of Medicine.

Though she does not possess the trait herself, a female who carries one of the recessive alleles can pass it on to her offspring. She is referred to as the trait is carrier. As there is a 50% chance that her sons will receive the X chromosome containing the color-blindness allele, half of them should be expected to carry the trait. The sons who inherit the recessive allele will manifest it because they only have one X chromosome. All of the woman's daughters will not have the trait, though, if the father is unaffected. The daughters will not experience colorblindness because the allele is recessive to the normal allele they inherited from their father, but they do have a 50% chance of inheriting the X chromosome containing it.

Another example of a recessive X-linked Mendelian trait is hemophilia, which is a disorder characterized by the blood's inability to clot normally. England's Queen Victoria carried the disorder. Two of her five daughters inherited the hemophilia allele from their mother and were carriers. When they married royalty in other European countries, they spread the allele across Europe, including to the royal families of Spain, Germany, and Russia. Victoria's son Prince Leopold also inherited the hemophilia allele from his mother, and he actually suffered from the disease. Understandably, hemophilia was once popularly called "the royal disease."

3.1.1.4. Feature: My Human Body

If you have this X-linked recessive disorder, you may be color blind or suspect that you may be; in any case, you perceive a world in which there are no discernible differences between any two colors. Some people who are color blind are not even aware that they are color blind; it is all they have ever known. Easy tests have been developed to identify the degree of color blindness in an individual. If you can clearly see the number 74 when you look at this circle, your red-green color vision is most likely normal. You can be red-green color blind if you are unable to see the number (Figure 3.6).

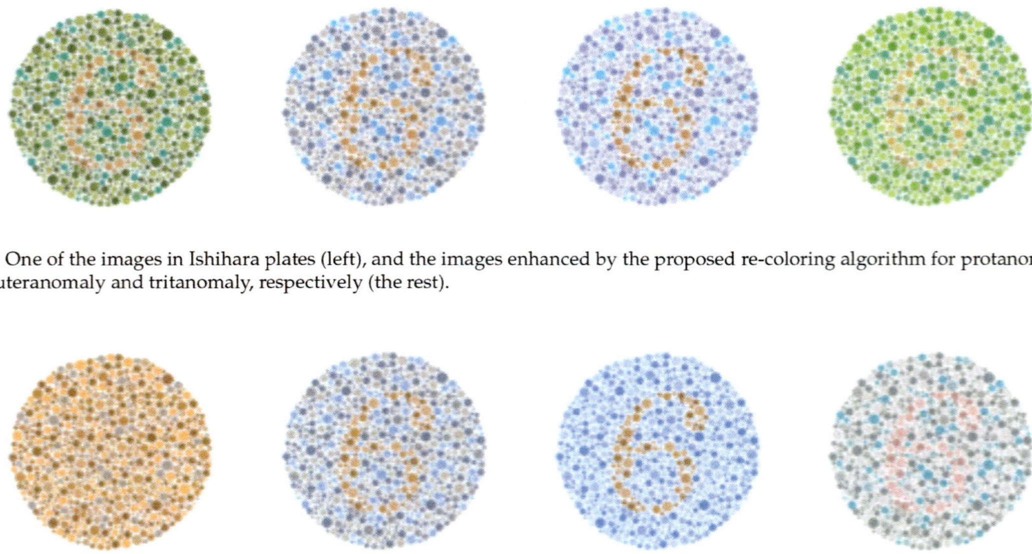

(**a**) One of the images in Ishihara plates (left), and the images enhanced by the proposed re-coloring algorithm for protanomaly, deuteranomaly and tritanomaly, respectively (the rest).

(**b**) The images in (a) generated by color vision deficiency simulation. The first image is deuteranomaly simulation of the original Ishihara plate. The rest images are the simulation results of protanomaly, deuteranomaly and tritanomaly on the re-colored images, respectively.

Figure 3.6. *Perception of colors.*

Source: Lin, H. Y., Chen, L. Q., & Wang, M. L., (2019). Improving discrimination in color vision deficiency by image re-coloring. Sensors, 19(10), 2250. https://doi.org/10.3390/s19102250.

Being color blind can cause a number of problems. These may range from minor frustrations to outright dangers. Here are a few examples:

- It could be challenging to color-coordinate furniture and clothes if you are color blind. It is possible that you'll wear color combinations that seem strange or clashing to someone with normal color vision.
- There are a lot of red or green LED indicator lights. To indicate whether they are on (green) or off (red), power strips and electronic devices, for instance, may have indicator lights.
- Color-coded test strips are frequently used for testing for common conditions like pH, hard water, and chemicals in swimming pools. For example, litmus paper used to measure pH turns red when an acid is present, but if you are color blind, you might not be able to see the test results.
- Do you prefer your steak well done? If you are color blind, you might not be able to distinguish between perfectly grilled or still undercooked (red) meat. It is also possible that you won't be able to tell ripe (red) from unripe (green) foods, like tomatoes. Furthermore, some foods—like dark green spinach—may

appear more like mud than food, which makes them utterly unappealing.

- Weather maps are frequently colored-coded. If you can't tell the difference between the two types of precipitation—green rain in your forecast and pink or red freezing rain—you might venture out onto the roads when you shouldn't and endanger yourself.
- Knowing the difference between a red and green traffic light could mean the difference between life and death. A person who is red-green color blind may find this extremely challenging. This kind of vision impairment prevents some people from being able to drive (Figure 3.7).

(a) Original (b) Protanopia

(c) Deuteranopia (d) Tritanopia

Figure 3.7. *Examples of challenges faced by those who are color blind.*

Source: Lin, H. Y., Chen, L. Q., & Wang, M. L., (2019). Improving discrimination in color vision deficiency by image re-coloring. Sensors, 19(10), 2250. https://doi.org/10.3390/s19102250.

3.2. NON-MENDELIAN INHERITANCE

Learning Objectives

- Discuss about multiple allele traits.
- Explain the feature: my human body

The non-Mendelian inheritance is characterized by characteristics that defy the Mendelian inheritance. Genes having multiple alleles are typically inherited in non-Mendelian patterns. True dominance or repression is not demonstrated by multiple alleles. The environment has a major influence on the offspring's phenotype. Nearly every characteristic in humans is inherited non-Mendelian. Multiple alleles, incomplete dominance, codominance, polygenic inheritance, phenotypic plasticity, and sex-linked traits are a few instances of non-Mendelian inheritance (Figure 3.8).

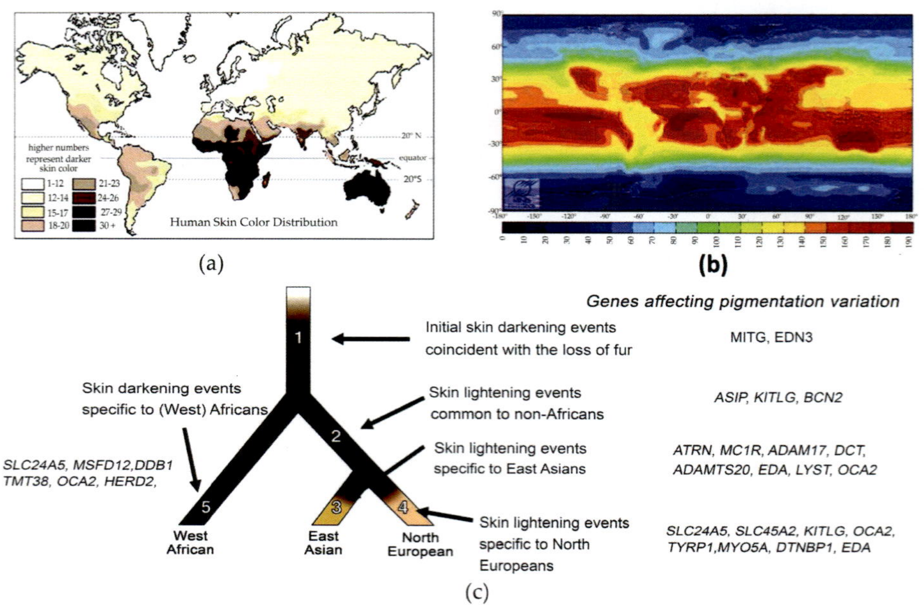

Figure 3.8. *Genetics of skin color.*

Source: Del Bino, S., Duval, C., & Bernerd, F., (2018). Clinical and biological characterization of skin pigmentation diversity and its consequences on UV impact. International Journal of Molecular Sciences, 19(9), 2668. https://doi.org/10.3390/ijms19092668.

The range of human skin tones, from very light to very dark, and all possible gradations in between, is depicted in this collage. As one might anticipate, the genetic basis of the skin color trait is more complex than that of a simple trait studied by Mendel in pea plants, which is based on a single gene with two alleles. Many other human characteristics, such as skin color, have more complex inheritance patterns than Mendelian traits.

3.2.1. Multiple Allele Traits

It is believed that most human genes contain more than two normal copies or alleles. Multiple allele traits are those governed by a single gene that have more than two alleles. ABO blood type is one illustration. A few proteins known as antigens are present on your red blood cells determines your blood type. The letters A, B, and O stand for the three common alleles for this characteristic (Figure 3.9).

Remember

These types of inheritance are referred to as non-Mendelian inheritance, and they include, among other things, the inheritance of traits with multiple alleles, codominance or incomplete dominance, and polygenic traits.

Blood Type

	A	B	AB	O
Red Blood Cell Type	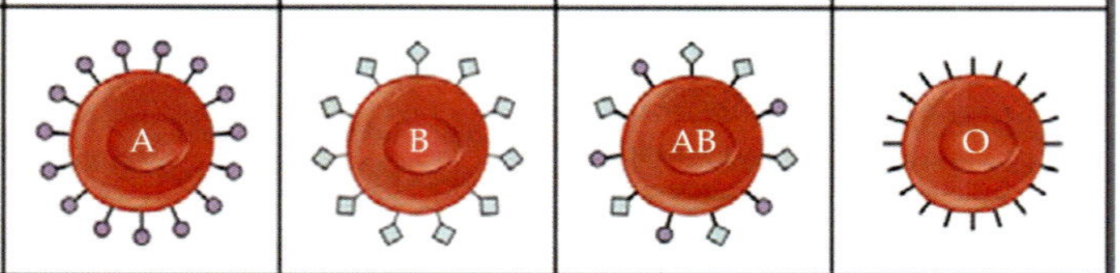 A	B	AB	O
Antibodies in Plasma	Cell Type	Anti-A	None	Anti-A and Anti B
Antigens in Red blood Cell	A antigen	B antigen	A and B antigens	None
Blood Types Compatible in an Emergency	A, O	B, O	A, B, AB, O (AB+ is the universal recipient)	O (O is the universal donor)

Figure 3.9. *ABO blood types per genotype.*

Source: https://courses.lumenlearning.com/suny-ap2/chapter/blood-typing/.

The three alleles, taken two at a time, result in six possible combinations, which is why there are six possible ABO genotypes. The O allele is subordinated to the A and B alleles. Because of this, people with the AA and AO genotypes exhibit the same phenotype, with type A blood containing the A antigen. The B antigen is present in the blood of people with the BB and BO genotypes, who share the same phenotype (type B blood). People with the OO genotype have no antigens for the ABO blood type (type O blood), meaning that no antigen is linked to the O allele.

3.2.1.1. Codominance

Analyze the ABO blood group table's genotype AB entry. With regard to the ABO blood type, alleles A and B are neither dominant nor recessive against one another. Codominant is the better term. When two gene alleles exhibit equal expression in the phenotype of heterozygotes, this phenomenon is known as codominance. Blood that has both A and B antigens in it is known as type AB blood, and individuals who are ABO heterozygotes exhibit this distinct phenotype.

Keyword

Genotype refers to the genetic makeup of an organism.

3.2.1.2. Incomplete Dominance

Incomplete dominance is a relationship that can also exist between alleles of the same gene. When the dominant allele is not totally dominant, this happens. Here, heterozygotes who inherit both alleles have an intermediate phenotype. This typically occurs when a gene's two alleles both produce proteins, but only one of those proteins functions. Consequently, the heterozygote person only produces half as much normal protein as a person who is homozygous for the normal allele.

A case of incomplete dominance in humans is the illness known as Tay Sachs. An enzyme that breaks down lipids is produced by the gene's normal allele in this instance. An enzyme that isn't functional is produced when an allele of the gene is faulty. It is sufficient for normal development for heterozygotes with one normal and one defective allele to produce half as much functional enzyme as a normal homozygote. However, homozygotes with only one faulty allele only produce inactive enzyme. This results in lipid buildup in the brain that begins in utero and seriously damages the brain. The majority of Tay Sachs disease sufferers pass away very young, usually before the age of five (Figure 3.10).

Hair type in humans is another excellent illustration of incomplete dominance. People who are heterozygous for one of the genes causing straight or curly hair will usually have wavy hair as their phenotype.

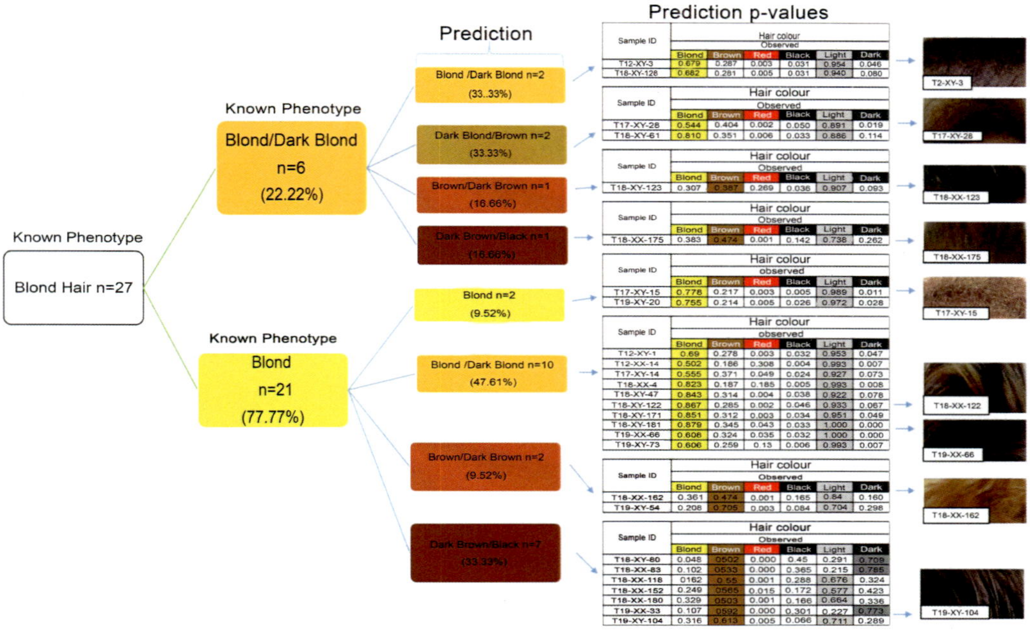

Figure 3.10. *Three phenotypes of hair through the incomplete dominance model.*

Source: Sari, O. I., Simsek, S. Z., Filoglu, G., & Bulbul, O., (2022). Predicting eye and hair color in a Turkish population using the HIrisPlex system. Genes, 13(11), 2094. https://doi.org/10.3390/genes13112094.

3.2.1.3. Polygenic Traits

Multiple genes are involved in the control of many human traits. We refer to these features as polygenic traits. Each gene's alleles contribute slightly to the phenotype. Particularly when each gene contains multiple alleles, there are a plethora of potential allele combinations. As a result, a wide range of phenotypes is feasible (Figure 3.11).

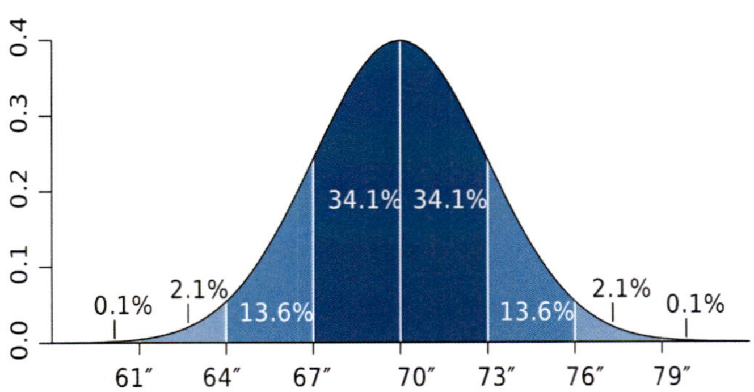

Figure 3.11. *Human adult height. Like many other polygenic traits, adult height has a bell-shaped distribution.*

Source: https://en.wikipedia.org/wiki/File:Standard_deviation_diagram_US_men_heights.svg.

An example of a human polygenic trait is adult height. Several genes, each with more than one allele, contribute to this trait, so there are many possible adult heights. One adult's height might be 1.655 m (5.430 feet), and another adult's height might be 1.656 m (5.433 feet), as shown in Figure 3.11. Adult height ranges from less than 5 feet to more than 6 feet, with males, on average, being somewhat taller than females. The majority of people fall near the middle of the range of heights for their se.

Did you know?

In 1928, Frederick Griffith discovered the phenomenon of transformation: dead bacteria could transfer genetic material to "transform" other still-living bacteria.

3.2.1.4. Environmental Effects on Phenotype

Environmental factors influence many traits in addition to genetic factors. For polygenic traits, this might be particularly true. Illness in childhood or poor nutrition, for instance, may have a negative effect on adult height. One more polygenic characteristic is skin color. People around the world have a vast variety of skin tones. In addition to differences in genes, differences in exposure to ultraviolet (UV) light cause some variation.

3.2.1.5. Pleiotropy

Multiple phenotypic traits are influenced by a single gene. Pleiotropy is the term for this. Pleiotropy in human beings is widely observed. They typically involve significant proteins that are necessary for multiple organ systems to develop or function normally. The gene responsible for the primary protein in collagen, a material that aids in the formation of bones, is an example of pleiotropy in humans. Additionally crucial to the eyes and ears is this protein. The gene's pleiotropic effects were found to be caused by mutations that cause issues with these sensory organs in addition to bones.

Pleiotropy can also be seen in sickle cell anemia. A mutation in the gene that typically codes for the red blood cell protein known as hemoglobin causes this recessive genetic disorder. Sickle cell hemoglobin, which is named for the sickle shape (shown in Figure 3.12) that the red blood cells take on in response to specific stimuli (such as physical exertion), is carried by two alleles in individuals with the disorder. The sickle-shaped red blood cells obstruct tiny blood vessels, which results in a variety of phenotypic symptoms such as strokes, kidney failure, deformities in the bones, and growth retardation.

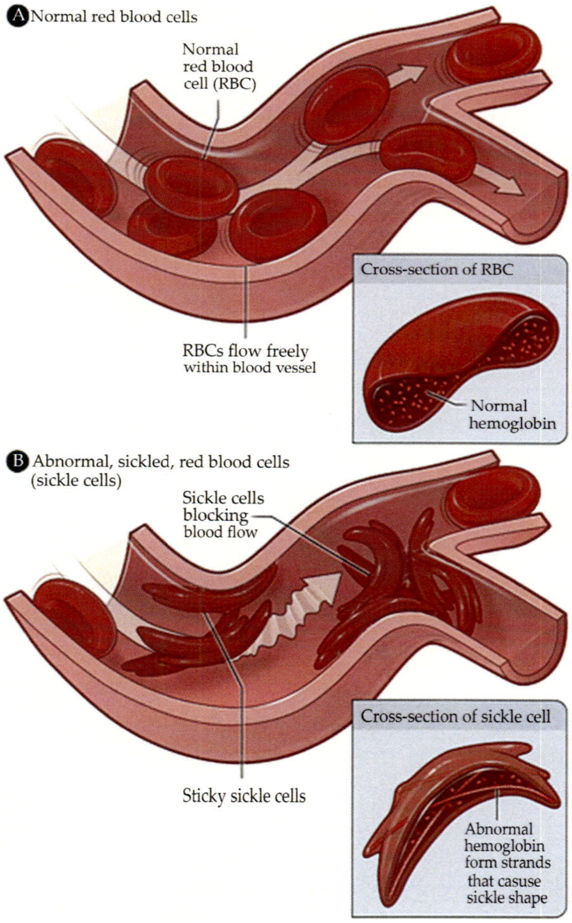

Figure 3.12. *For comparison, the sickle-shaped red blood cell on below is shown next to several normal red blood cells.*

Source: By The National Heart, Lung, and Blood Institute (NHLBI) – http://www.nhlbi.nih.gov/health/health-topics/topics/sca/, Public Domain, https://commons.wikimedia.org/w/index.php?curid=19198765.

3.2.1.6. Epistasis

Gene expression is influenced by certain genes. We refer to this as epistasis. Similar to dominance, epistasis happens between distinct genes as opposed to different alleles of the same gene. Epistasis can be seen in albinism, for example. The skin pigmentation of an albino is almost nonexistent. In contrast to the genes that determine skin color, a completely different gene causes the condition. The cause of albinism is a gene mutation that prevents the production of tyrosinase, a protein required for the synthesis of normal skin pigment. Irrespective of inherited skin color genes, an individual with the albinism mutation will lack all skin pigment.

3.2.2. Feature: My Human Body

Do you know what your ABO blood type is? This vital information might just save your life in an emergency situation. Making sure the blood you receive matches your blood type is crucial if you ever need a transfusion. The explanation for this is that if the blood transfusion contains an antigen that is absent from your own blood, the antibodies in your blood plasma—the liquid portion of your blood—will identify the antigen as alien to your body and trigger an agglutination reaction. Red blood cells that have been transfused will aggregate in this response. A dangerous and possibly lethal agglutination reaction occurs.

It will be easier for you to determine which type(s) of blood you can safely receive if you ever need a transfusion if you are aware of the antigens and antibodies found in each of the ABO blood types. This data is displayed for each ABO blood type in (Figure 3.13). If you have blood type A, anti-B antibodies are present in your blood plasma and your red blood cells carry the A antigen. Your anti-B antibodies would attack the transfusion-provided red blood cells, leading to agglutination, if you were to receive a transfusion of either type B or type AB blood, which both have the B antigen.

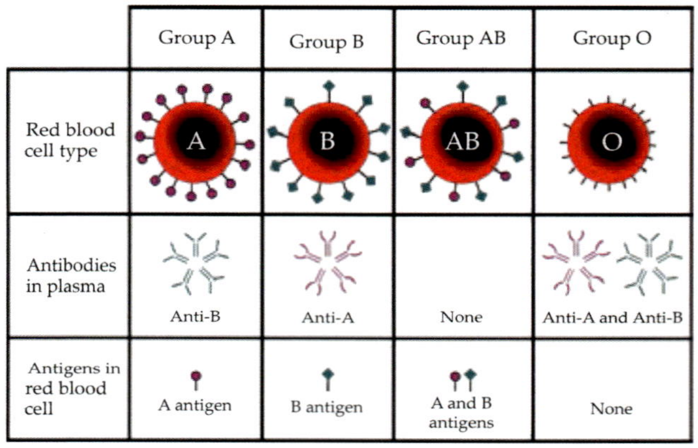

Figure 3.13. *Antigens and antibodies in ABO blood types.*

Source: https://upload.wikimedia.org/wikipedia/commons/thumb/3/32/ABO_blood_type.svg/1200px-ABO_blood_type.svg.png.

It is possible that you have heard the terms "universal donors" and "universal recipients" used to describe people with blood type O and type AB, respectively. Individuals with type O blood can donate blood to anyone because they do not contain the A or B antigens, which prevents an immune reaction when their blood is transfused into someone with a different ABO blood type. Individuals with type AB blood, on the other hand, are able to receive a blood transfusion from anyone because they do not contain anti-A or anti-B antibodies. Which blood type or blood types can safely receive a transfusion of type AB blood, and which blood type or blood types can those with type O blood safely receive?

The two approaches that explain the genetic basis of phenotypes in heterozygous individuals for a given trait are Mendelian and non-Mendelian inheritance. The inheritance of phenotypes that are solely based on two alleles is referred to as Mendelian inheritance. For the phenotype, one of the two alleles is dominant and the other is recessive. The process by which several alleles and polygenes interact to determine phenotypes is known as non-Mendelian inheritance. Thus, the influence of the number of alleles or genes involved in the determination of a particular trait is the main distinction between Mendelian and non-Mendelian inheritance.

Did you get it?

1. **What do you understand by codominance?**
2. **Explain the polygenic traits?**

3.3. MULTIPLE FACTOR INHERITANCE

Learning Objectives

- Define the methods of multiple factor inheritance

Genes can be categorized as "major genes" if they have a significant impact, or as "minor genes" if they have a tiny impact on their own. The division of genes into major and minor categories is somewhat arbitrary, as the impact of individual allelic variations on phenotypic variation can vary greatly. Nevertheless, it is helpful to ascertain whether allelic variations at one or two loci or multiple loci account for the genotypic variance of a trait. For example, the body weight of mice is a characteristic in which the genetic variation is usually due to segregation and assortment of alleles with minor effects at many loci. However, in addition to the genetic causes of normal phenotypic variation, there are mutations with drastic major effects on body weight such as the obese gene (ob) and the pituitary dwarf gene (dw).

If a characteristic's genetic variation is influenced by allelic differences at multiple loci rather than just a few, it is said to exhibit multiple factor inheritance. Many "minor genes" work together to cause multiple factor inheritance. Instead of defining the issue in terms of the mode of inheritance, Falconer defined it in terms of the character itself. "Any attribute for which individual differences do not divide the individuals into qualitatively distinct classes," is how he defined a quantitative character.

According to this definition, the number of loci pertaining to a trait is genetic variation is irrelevant as long as the segregation of individual alleles does not result in a distinct phenotypic discontinuity. Regarding genetic variation, it is generally convenient to categorize it into two types: one or a few loci vs. numerous loci and the two types of phenotypic variation (discrete vs. continuous). Analytical techniques suitable for quantitative characters are needed for all four combinations of genetic variation and phenotypic variation, with the exception of one or a few loci influencing variation in a discrete trait.

3.3.1. Methods of Multiple Factor Inheritance

We assume that numerous allelic differences with negligible individual effects on a trait contribute to the phenotypic variation of a quantitative trait, at least partially. Although it is impossible to identify and investigate the mode of inheritance of every allele causing phenotypic variation, it is possible to assess the relative contributions of genetic and environmental factors. Although a trait is genetic basis may be indicated by the resemblance between relatives, which may also result from shared environmental factors. This group includes cage effects and maternal effects. It is possible to estimate the genetic and environmental factors and establish the relative importance of each by employing suitable experimental designs and analyzes. For these kinds of research, inbred mouse strains are highly helpful.

Keyword

Phenotype is the set of observable characteristics or traits of an organism.

The terms "genetic determination" and "heritability" are often used interchangeably. Genetic determination will mean that portion of the total phenotypic variation attributable to any genetic causes, and heritability denoted as h^2, will mean that portion of the total phenotypic variance attributable to additive genetic effects alone. Heritability can therefore be a part of genetic determination.

1. **Comparing Two Strains:** The degree of genetic determination can be estimated by comparing two inbred strains, their F_1, F_2, and backcross generations. Under the most simplifying assumptions, the phenotypic variance (σ_P^2) may be considered as the sum of the genetic variance component (σ_G^2) and the environmental variance component (σ_E^2): $\sigma_P^2 = \sigma_G^2 + \sigma_E^2$. If the parental strains are highly inbred, there will be only environmental and no detectable genetic variance within the two parental strains and the F_1 hybrids. In the F_2 animals, the phenotypic variance will be ascribable only in part to environmental variance. In addition, there will be a genetic component of variance due to the segregation of alleles and the assortment of nonalleles. Therefore, one can estimate genetic determination by subtracting the F_1 computed phenotypic variance (V_{F1}, environmental) from the F_2 computed phenotypic variance (V_{F2}, genetic and environmental) and dividing by the F_2 computed phenotypic variance:

$$\text{Genetic determination} = \frac{VF_2 - VF_1}{VF_2} = \frac{\left(\sigma G^2 + \sigma E^2\right) - \sigma E^2}{\left(\sigma G^2 + \sigma E^2\right)} = \frac{\sigma G^2}{\sigma P^2}$$

Grahn used this method in estimating the genetic determination of radiation resistance in mice. An approximation of the minimum number of loci affecting a character is often of interest. This can be estimated from the equation:

$$n = \frac{(m_1 - m_2)^2}{8\,(V_{F_2} - V_{F_1})}$$

Where; m_1 and m_2 are the means of the parental strains; and V_{F2} and V_{F1} are the computed variances of the F_2 and F_1 generations. This is considered as a minimum estimate because the equation is based on a number of assumptions. Violation of these assumptions will tend to underestimate *n*.

2. **Comparing Many Strains:** Variations among strains have been identified for so many characters that one could almost take genetic influences on character variation for granted. By comparing numerous strains, one can estimate the relative importance of genetic and environmental factors; the reliability of the estimate increases with the number of unrelated inbred strains used in the analysis. The process involves characterizing multiple highly inbred strains and using standard analysis-of-variance techniques to estimate the within- and between-strain variance components. The expected mean square between strains is equal to $\sigma_E^2 + k(1 + F)\sigma_G^2$, where F is the inbreeding coefficient, σ_E^2 and σ_G^2 are the environmental and genetic components of variance respectively, and k is the number of individuals measured per strain (or the harmonic mean of the numbers in the strains if the sample sizes are not equal). F is effectively equal to 1 in highly inbred strains, so the genetic variance component is doubled.

3. **Comparing Inbred and Randombred Strains:** By comparing the estimate with the variance of the segregating F generation, one can ascertain the extent of genetic influence on a character. The methods above explain how nongenetic variance can be estimated from the phenotypic variance within inbred strains and within their F_1 hybrids. By comparing inbred and Randombred strains, this technique can also be used to determine a character's level of genetic determination. Falconer and Bloom used six inbred strains and their 15 F_1 hybrids to estimate the nongenetic variance in the incidence of induced lung tumors in mice. Using this estimate, they partitioned the variance in two random bred lines (JC and LX) into genetic and environmental components and showed that individual genetic differences account for 82% of the variation of susceptibility in the JC strain and 88% in the LX strain.

4. **Diallel Cross:** Although it can be used to estimate both additive and nonadditive components of genetic variance, the diallelic cross technique has not been widely applied in mouse studies about quantitative traits. The method basically entails generating every possible cross between a number of inbred strains and calculating each line's general combining ability, as well as the specific combining ability of each line and each pair of strains. The general combining ability of a strain is represented by its average performance, and the specific combining ability of a cross is represented by the deviation in performance from the average combining ability of the two strains. The nonadditive genetic variance can be attributed to the specific combining ability component, which is comparable to an interaction term

in a two-way analysis of variance. Additive genetic variance accounts for variations in overall combining ability.

5. **Selection:** The definition of selection provided by Lerner is "the non-random differential reproduction of genotypes." Natural selection results in a change in gene frequencies due to often obscure and complex natural causes. Natural selection operates based on the fitness characteristic, which has inherent complexities and obscurities. Artificial selection refers to deliberate human efforts to cause a variation in the spread of genotypes. "Artificial" does not mean that human-chosen selection isn't real selection. Research on natural selection in mice has mostly focused on how it affects a specific mutant that has been incorporated into the wild population. Who gave a t allele to a mouse population on an island? The two main uses of artificial selection in mice are to investigate the genetic theory of selection (e.g., Falconer's selection for tail length) and to generate a significant divergence in a trait to investigate the physiological genetics of the trait (e.g., Wolfe's selection for blood pH).

   Only when additive genetic variance affects the trait under consideration can selection change the phenotypic mean of the population. Inbred strains are not appropriate for a selection experiment because they exhibit no discernible genetic variability. But by crossing two, three, four, or more inbred strains with genetically distinct origins, genetically heterogeneous foundation populations can be created. To create a foundation population for selection, inbred strains crossed in a particular way are used. This has the added benefit of making the foundation population replicable at a different time and location.

Keyword

Genetic Variance – It involves the inheritance of a particular allele from your parent and this allele's independent effect on the specific phenotype, which will cause the phenotype deviation from the mean phenotype.

The higher the proportion of additive genetic variance relative to the phenotypic variance (that is, the greater the heritability, h^2) the greater will be the gains under selection. Falconer gave ways of estimating h^2 prior to selection by using the regression of the means of offspring on the means of parents and by paternal half sib analysis. Selection based on the phenotype of the individual is most efficient under most circumstances, provided that there is a reasonable level of heritability. Falconer provided an illustration of a phenotypic-based selection strategy. Every generation uses 10 mated pairs, and the two most extreme individuals from each of the 10 litters are chosen to be the next generation's individuals. After that, these 20 people mate at random with each other, avoiding sibling pairings. This system works well because it uses a manageable number of mice per generation and maintains a low level of inbreeding over a number of generations. This method does not allow selecting for maternal effects under genetic control, which may affect the trait, but it does avoid selecting for environmental variation due to common litter or cage effects because it is within-litter selection.

Did you get it?

1. **Discuss the comparing inbred and Randombred strains**
2. **What do you understand by diallel cross?**

FOCUS ON CAREERS

Genetic Counsellor: Genetic counselling is a specialized healthcare profession that focuses on helping individuals and families understand and adapt to the medical, psychological, and familial implications of genetic conditions. Genetic counsellors provide education, support, and guidance to individuals considering genetic testing, facing a diagnosis of a genetic disorder, or navigating reproductive decision-making. Here's a closer look at the career of a genetic counselor:

Role and Responsibilities: Genetic counselors work with individuals and families to:

- Assess personal and family medical histories to identify potential genetic risks.
- Educate clients about genetic inheritance patterns, genetic testing options, and disease management strategies.
- Provide emotional support, empathy, and counseling to help clients cope with the psychological impact of genetic conditions.
- Interpret genetic test results, explain implications, and discuss implications for family members.
- Assist clients in making informed decisions about medical treatments, reproductive options, and family planning.
- Collaborate with other healthcare professionals, including physicians, nurses, and laboratory scientists, to coordinate patient care and ensure comprehensive support.

Settings and Employment Opportunities: Genetic counselors work in various healthcare settings, including:

- Hospitals and medical centers: Providing genetic counseling services in departments such as medical genetics, oncology, obstetrics and gynecology, and pediatrics.
- Specialty clinics: Focusing on specific areas of genetic counseling, such as cancer genetics, prenatal genetics, cardiovascular genetics, or neurogenetics.
- Genetic testing laboratories: Offering pre-test and post-test counseling, result interpretation, and support services for individuals undergoing genetic testing.
- Research institutions: Conducting research on genetic counseling practices, genetic testing technologies, and psychosocial aspects of genetic conditions.

Education and Training: To become a genetic counselor, individuals typically complete the following steps:

- Earn a bachelor's degree in a relevant field such as biology, genetics, psychology, or counseling.

- Obtain a master's degree in genetic counseling from an accredited genetic counseling program, which includes coursework in genetics, counseling techniques, medical ethics, and clinical rotations.
- Complete supervised clinical training through internships or practicum experiences in genetic counseling settings.
- Obtain certification as a genetic counselor by passing the American Board of Genetic Counselling (ABGC) certification exam.

Professional Development and Continuing Education:

Genetic counselors engage in ongoing professional development to stay abreast of advances in genetics and counseling practice. This may include attending conferences, participating in continuing education programs, joining professional organizations such as the National Society of Genetic Counselors (NSGC), and pursuing advanced certifications or specialized training in specific areas of genetic counseling.

- **Personal Qualities and Skills:** Successful genetic counselors possess a blend of clinical expertise, communication skills, and interpersonal qualities, including:
  - Empathy, compassion, and sensitivity to the emotional needs of clients.
  - Strong communication and counseling skills to convey complex genetic information in clear, understandable terms.
  - Critical thinking and problem-solving abilities to assess genetic risks, interpret test results, and facilitate decision-making.
  - Cultural competence and awareness of diversity issues to provide culturally sensitive care to clients from diverse backgrounds.
  - Ethical integrity and professionalism to uphold the highest standards of patient care and confidentiality.

A career as a genetic counselor offers a rewarding opportunity to make a meaningful difference in the lives of individuals and families affected by genetic conditions, empowering them to make informed decisions about their health and well-being.

ROLE MODEL

GREGOR MENDEL: *FATHER OF MODERN GENETICS*

Gregor Mendel, born Johann Mendel, was an Austrian scientist and monk hailed as the "Father of modern genetics" for his pioneering research in the field of heredity. He was a monk in Augustinian Abbey of St Thomas in Brno where he worked as a teacher. He had a deep interest in botany which led him to conduct experiments on pea plants. Inspired by the work of a biologist named Franz Unger, he began his experiments in the monastery's sprawling gardens. Over the course of his study, he observed that there were seven characteristics in the pea plants and two forms of each characteristic. These characteristics included seed shape and pod shape in addition to plant height and seed color. Mendel observed that the seven characteristics he had recognized remained consistent over generations in purebred plants. For eight years, he carefully crossbred and grew thousands of pea plants, and patiently analyzed and compared the plants and seeds for the difference in color and size of the seeds, and variations in length of the plants. He took various precautions to prevent the accidental pollination of the flowers which could have altered the results of the experiments. His meticulous study and the resultant observations led to what is today known as Mendel's Laws of Inheritance.

Childhood and Early Life:

- Gregor Mendel was born as the middle child and only son of Anton and Rosine Mendel. He had two sisters and the family lived and worked on the farm they had owned for generations.
- As a child, he worked in the garden and studied beekeeping which cultivated in him a deep love for biological sciences.
- He received his early schooling in his own small village but had to be sent to a nearby town for his secondary education. The decision to send away their only son was not an easy one for his parents, but they did it for the sake of his future.
- Later on, he went to the University of Olomouc where he studied philosophy and physics from 1840 to 1843.

Career and Works:

- In 1843, he began his training as a priest and joined the Augustinian Abbey of St Thomas in Brno as a monk. He took the name 'Gregor' on entering the religious field.
- The monastery sent him to the University of Vienna to study under Abbot C.F. Napp. There he studied physics and mathematics under Christian Doppler and botany under Franz Unger.
- He rejoined the monastery as a teacher in 1853 where he was motivated by his colleagues to conduct a study on plants.
- He began to conduct his practical study on plants in 1856. He studied edible pea plants and recognized seven distinct characteristics that remained consistent over generations in purebred varieties. These characteristics included: the height of the plant, the shape of the pod, the shape of the seed, size and color of the seeds, etc.
- He cross-pollinated the plants with contrasting characteristics in order to study the effects on the offspring. He also took due precaution to prevent accidental pollination by insects. He cultivated thousands of pea plants over the course of his experiments.
- He collected the seeds of the offspring and analyzed them for variations in color, shape, and size. He also compared the plants for differences in height.
- Over a period of eight years, he painstakingly examined the plants, pods and seeds and made observations that would form the basis for a deeper study of genetics.
- He presented the results of his experiments at the Natural History Society of Brno in 1865. His findings were published in a paper 'Experiments on Plant Hybridization' in 1866. However, his research failed to create an impact at that time.
- In 1868, he was made abbot of the monastery where he had been teaching for the past many years. The increased responsibilities prevented him from conducting any further scientific experiments.
- Gregor Mendel's works failed to gain much importance during his lifetime but formed the foundation for what is today known as Mendel's Laws of Inheritance.

Major Works:

- Mendel through his extensive experimentation and analysis founded the three laws or principles of inheritance: The law of segregation, the law of dominance, and the law of independent assortment.
- He developed the concepts of dominant and recessive genes that explain how genetic traits are passed along from generation to generation.

- His 1865 paper 'Experiments on Plant Hybridization' which was largely ignored during his lifetime is today regarded as the base of genetic experimentation.

Personal Life and Legacy:

- As a young man, he had very close and loving relations with his parents. Being a monk, he never married and led a life of celibacy.
- He died at the age of 61 after suffering from kidney problems.
- His work on heredity which did not find much acceptance during his lifetime took on much greater significance after his death and he was posthumously hailed as the father of modern genetics.

REFERENCES

1. Frank, S. A. (2004). Genetic predisposition to cancer–Insights from population genetics. Nature Reviews Genetics, 5(10), 764–772.
2. Griffiths, A. J., Miller, J. H., Suzuki, D. T., Lewontin, R. C., & Gelbart. (2000). Genetics and the organism: Introduction. An Introduction to Genetic Analysis (7th ed.). New York: W.H. Freeman.
3. Hartl, D., & Jones, E. (2005). Genetics: Analysis of Genes and Genomes (6th ed.). Jones & Bartlett.
4. King, R. C., Mulligan, P. K., & Stansfield, W. D. (2013). A Dictionary of Genetics (8th ed.). New York: Oxford University Press.
5. Poczai, P., Santiago-Blay, J. A., Sekerák, J., Bariska, I., & Szabó, A. T. (2022). Mimush sheep and the spectre of inbreeding: Historical background for Festetics's organic and genetic laws four decades before Mendel's experiments in peas. Journal of the History of Biology, 55(3), 495–536.
6. Staff. (2005). Gene therapy (FAQ). Human Genome Project Information. Oak Ridge National Laboratory.

CHAPTER 4

Gene Expression and Regulation

LEARNING OBJECTIVES

After studying this chapter, you will be able to:

- Discuss the regulation of gene expression;
- Focus on transcriptional and post-transcriptional control
- Discuss the translational and post-translational control
- Explain the epigenetics: mechanisms and implications.

INTRODUCTORY EXAMPLE

Fundamental Introductory of Gene Expression

Gene expression is the process by which information encoded in the DNA of a gene is used to direct the synthesis of a functional gene product, typically a protein. This intricate molecular dance occurs within the cells of all living organisms and plays a fundamental role in determining the characteristics and behaviors of organisms. To grasp the concept of gene expression, let's delve into a hypothetical scenario involving a group of cells in a plant undergoing environmental stress.

Imagine a lush meadow where a species of flowering plants, let's call them "Sunflowersia," thrives under the warm sun. These sunflowers possess genes responsible for producing enzymes that facilitate photosynthesis, the process by which they convert sunlight into energy. The expression of these genes ensures the plants' ability to thrive in their sunny habitat.

However, one summer, an unexpected heatwave strikes the meadow, subjecting the Sunflowersia plants to intense temperatures. As a result of this environmental stress, the plants must adapt their gene expression patterns to cope with the challenging conditions.

At the molecular level, gene expression begins with the transcription of DNA into messenger RNA (mRNA) molecules by an enzyme called RNA polymerase. This process, known as transcription, occurs in the nucleus of eukaryotic cells, where the DNA resides. In our Sunflowersia plants, certain genes encoding heat shock proteins are activated in response to the heatwave. When the plants sense the elevated temperatures, specific transcription factors bind to regulatory sequences in the DNA, promoting the transcription of these heat-responsive genes into mRNA.

This is followed by the process of mRNA processing, during which the newly transcribed mRNA undergoes modifications before leaving the nucleus. These modifications, including the addition of a protective cap and a poly-A tail, help stabilize the mRNA and facilitate its export to the cytoplasm. In the cytoplasm, the mRNA serves as a template for protein synthesis during translation.

Translation occurs on ribosomes, complex molecular machines composed of proteins and RNA molecules, which read the mRNA's nucleotide sequence and assemble amino acids into a polypeptide chain. The sequence of amino acids in the polypeptide chain is determined by the sequence of codons—three-nucleotide sequences—on the mRNA. In our Sunflowersia plants, the newly synthesized heat shock proteins fold into specific three-dimensional shapes, enabling them to function as molecular chaperones that help other proteins maintain their proper structure under heat stress.

Once synthesized, the heat shock proteins and other proteins produced in response to the heatwave help the Sunflowersia plants withstand the adverse conditions. These proteins may assist in stabilizing damaged proteins, preventing protein aggregation, or facilitating the refolding of denatured proteins—all essential functions for cellular survival in extreme temperatures.

As the heatwave subsides and environmental conditions return to normal, the expression of heat shock protein genes gradually decreases, allowing the Sunflowersia plants to resume their typical gene expression profiles associated with optimal growth and development. However, the experience of the heatwave may leave a lasting imprint on the plants' gene expression patterns, influencing their responses to future environmental stresses.

UNIT INTRODUCTION

Gene expression is the process by which a gene's information is used to create a functional product, usually a functional RNA molecule or protein. The manifestation of traits in organisms and the control of cellular activities depend on this process. Transcription and translation are the two primary phases of gene expression.

The process of transcription, which is the initial stage of gene expression, involves the RNA polymerase enzyme copying a gene's DNA sequence into a corresponding RNA molecule. The messenger RNA (mRNA) molecule transports genetic information from the nucleus' DNA to the cytoplasm, where it functions as a template for the synthesis of proteins. The rate and specificity of RNA synthesis are determined by a number of factors and regulatory elements, making transcription a highly regulated process. These include chromatin remodeling complexes, which alter DNA's accessibility to the transcription machinery, and transcription factors, which bind to particular DNA sequences known as enhancers and promoters.

The mRNA molecule is processed after transcription, going through steps like capping, splicing, and polyadenylation to become a mature mRNA that ribosomes can translate. The second phase of gene expression, known as translation, involves decoding the mRNA's nucleotide sequence in order to put together a particular arrangement of amino acids to form a polypeptide chain. In addition to ribosomes, which catalyze the creation of peptide bonds between neighboring amino acids, this process also involves transfer RNA (tRNA) molecules, which are able to identify and bind to particular codons on the mRNA. To guarantee the exact control of protein synthesis in response to diverse internal and external stimuli, gene expression is strictly regulated at several levels. Transcriptional regulation, post-transcriptional regulation, translational regulation, and post-translational modification of proteins are examples of regulatory mechanisms. Through the use of these regulatory mechanisms, cells can modify their gene expression profiles to suit their developmental stage, physiological needs, and external environment, maintaining homeostasis and adapting to changing conditions. It is crucial to comprehend the mechanisms underlying gene expression and regulation because dysregulation of gene expression can result in a number of diseases, such as cancer, developmental disorders, and metabolic disorders.

4.1. REGULATION OF GENE EXPRESSION

Learning Objectives

- Differentiate between Prokaryotic and Eukaryotic Gene Expression
- Focus on Prokaryotic Gene Regulation
- Explain the Eukaryotic Gene Regulation

Gene regulation, also known as regulation of gene expression, refers to the variety of methods that cells employ to either up- or down-regulate the synthesis of particular gene products (proteins or RNA). Biology is replete with examples of complex gene expression programs that initiate developmental pathways, react to environmental cues, or help an organism adapt to new food sources. It is possible to control almost every stage of the expression process of a gene, including transcriptional initiation, RNA processing, and protein post-translational modification. In a gene regulatory network, one gene regulator frequently affects another, and so on.

Gene regulation is essential for viruses, prokaryotes and eukaryotes as it increases the versatility and adaptability of an organism by allowing the cell to express protein when needed. Gene regulation in multicellular organisms drives morphogenesis and cellular differentiation in the embryo, resulting in the formation of distinct cell types with varying gene expression profiles derived from the same genome sequence. This is fundamental to the study of evolutionary developmental biology and provides some insight into how evolution operates at the molecular level, even though it does not explain the origin of gene regulation.

The DNA of every somatic cell in the body is essentially the same. A few examples are mature red blood cells, which have no DNA, and certain immune system cells that rearrange their DNA during the production of antibodies. Even though the functions of the liver and eyes are very different, the genes that dictate things like how quickly you metabolize food or whether you have brown hair or green eyes are generally the same in both types of cells. Why do cells in the liver differ so much from those in the eye, and why do all cells have the same DNA, but differ so much in terms of structure and function?

Every cell in your body has the same DNA sequences, but not every cell expresses or turns on the same set of genes. Actually, each cell can only produce a small subset of all proteins. That is, not every gene encoded in a piece of DNA gets translated into mRNA or a protein in any given cell. One type of protein is produced by cells in the liver, while another type is produced by cells in the eye. Furthermore, liver cells may produce distinct subsets of liver proteins at different times. There are numerous levels and stages of regulation involved in the highly regulated process of gene expression. This intricacy guarantees that every protein is expressed in the right cells at the right time (Figure 4.1).

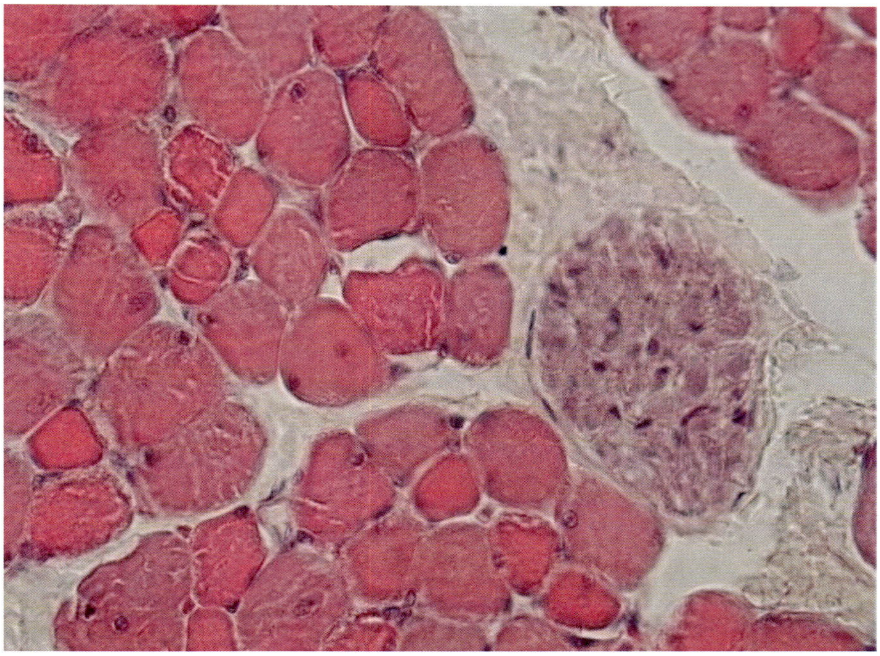

Figure 4.1. *The genetic content of each somatic cell in an organism is the same, but not all genes are expressed in every cell.*

Note: The control of which genes are expressed dictates what kind of cell it would be. Shown here is a cross-section through skeletal muscle and a small nerve at high magnification. It is the differential gene expression patterns that arise in different cells that give rise to a complete organism.

Source: CC BY-SA 3.0, https://commons.wikimedia.org/w/index.php?curid=657552.

The right timing of protein synthesis is essential for a cell's correct operation. All cells use information encoded in their DNA to regulate or control the synthesis of proteins. Gene expression is the process of "turning on" a gene to produce mRNA and protein. Within a basic unicellular organism or an intricate multicellular organism, every cell regulates the timing of gene expression, the amount of protein produced, and the decision to cease protein synthesis when it is no longer required.

Gene expression regulation preserves both space and energy. Requiring the genes to be activated only when necessary results in greater energy efficiency. Furthermore, because transcription and translation require the unwinding of DNA from its tightly coiled structure, only a subset of genes are expressed in each cell, saving space. If every protein was constantly expressed in every cell, then cells would need to be very large. The regulation of gene expression is incredibly intricate. Problems with this process are harmful to the cells and can cause a variety of illnesses, including cancer.

4.1.1. Deference Between Prokaryotic and Eukaryotic Gene Expression

Prokaryotic organisms are single-celled creatures without a cell nucleus; therefore, their DNA is free to move around in the cytoplasm of the cell. The gene encoding a specific protein is translated into mRNA, which is then translated into protein when the protein is required. Translation stops when the protein is no longer required. Therefore, transcription regulation is the main way to manage the amount of each protein expressed in a prokaryotic cell.

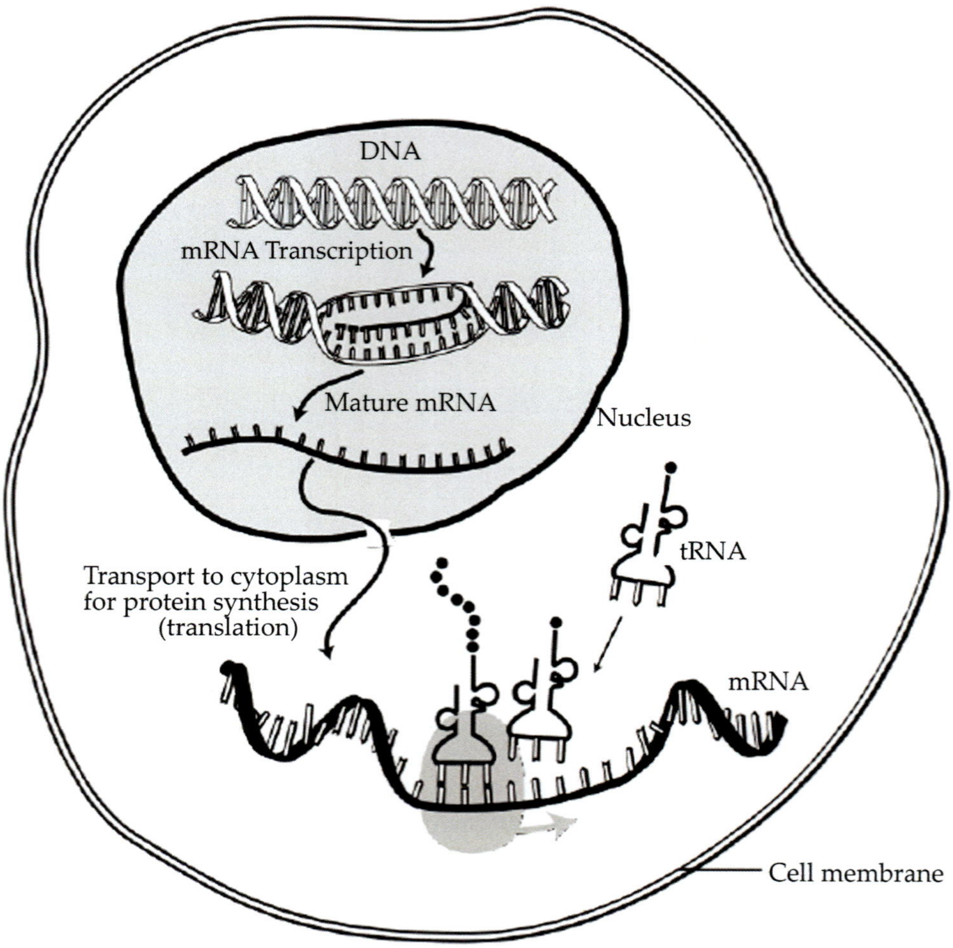

Figure 4.2. *Prokaryotic transcription and translation occur simultaneously in the cytoplasm, and regulation occurs at the level of transcription. In eukaryotes, transcription and translation are physically separated, and gene expression is regulated at many different levels.*

Source: By National Human Genome Research Institute, NIH – Own work using mRNA illustration, NIH, Public Domain, https://commons.wikimedia.org/w/index.php?curid=126381443.

In contrast, the complexity of eukaryotic cells is increased by intracellular organelles. The DNA of eukaryotic cells is located in the nucleus, where it is translated into mRNA. Ribosomes convert the freshly created mRNA into protein in the cytoplasm after it has been altered and moved from the nucleus. The nuclear membrane acts as a physical barrier between the transcription and translation processes, which are limited to the nucleus and the cytoplasm, respectively. In eukaryotes, there is potential for regulation of gene expression throughout the entire process (Figure 4.2).

Some of the differences in the regulation of gene expression between prokaryotes and eukaryotes are summarized in Table 4.1.

Table 4.1. Differences in Prokaryotic and Eukaryotic Gene Regulation

Prokaryotic Organisms	Eukaryotic Organisms
Lack nucleus.	Contain nucleus.
DNA is found in the cytoplasm.	DNA is in the nucleus.
Transcription and translation occur almost simultaneously.	Transcription occurs in the nucleus prior to translation, which occurs in the cytoplasm.
Gene expression is regulated primarily at the transcriptional level.	Gene expression is regulated at many levels: epigenetic, transcriptional, nuclear shuttling, post-transcriptional, translational, and post-translational.

Source: https://courses.lumenlearning.com/wm-biology1/chapter/reading-prokaryotic-and-eukaryotic-gene-regulation/.

4.1.2. Prokaryotic Gene Regulation

Prokaryotes have circular chromosomes made of arranged DNA that are found in the cytoplasm of the cell. Proteins that share a biochemical pathway or are required for a particular function are frequently encoded in groups known as operons.

For example, all five of the genes needed to make the amino acid tryptophan in the bacterium *E. coli* are located next to each other in the *trp* operon. The genes in an operon are transcribed into a single mRNA molecule. This allows the genes to be controlled as a unit: either all are expressed, or none is expressed. Each operon needs only one regulatory region, including a promoter, where RNA polymerase binds, and an operator, where other regulatory proteins bind.

There are three different kinds of regulatory molecules in prokaryotic cells that can influence operon expression. Proteins called activators boost a gene's transcription. Proteins called repressors prevent a gene from being transcribed. Inducers are molecules that bind to repressors and cause them to become inactive. Below are two examples of how these molecules regulate different operons.

4.1.2.1. The trp Operon: A Repressor Operon

Like all cells, bacteria need amino acids to survive. Tryptophan is one amino acid that the bacterium *E. coli* can either ingest from the environment or synthesize. When *E. coli* needs to synthesize tryptophan, it must express a set of five proteins that are encoded by five genes. These five genes are located next to each other in the tryptophan (*trp*) operon (Figure 4.3).

When tryptophan is present in the environment, *E. coli* does not need to synthesize it, and the *trp* operon is switched off. However, when tryptophan availability is low, the *trp* operon is turned on so that the genes are transcribed, the proteins are made, and tryptophan can be synthesized.

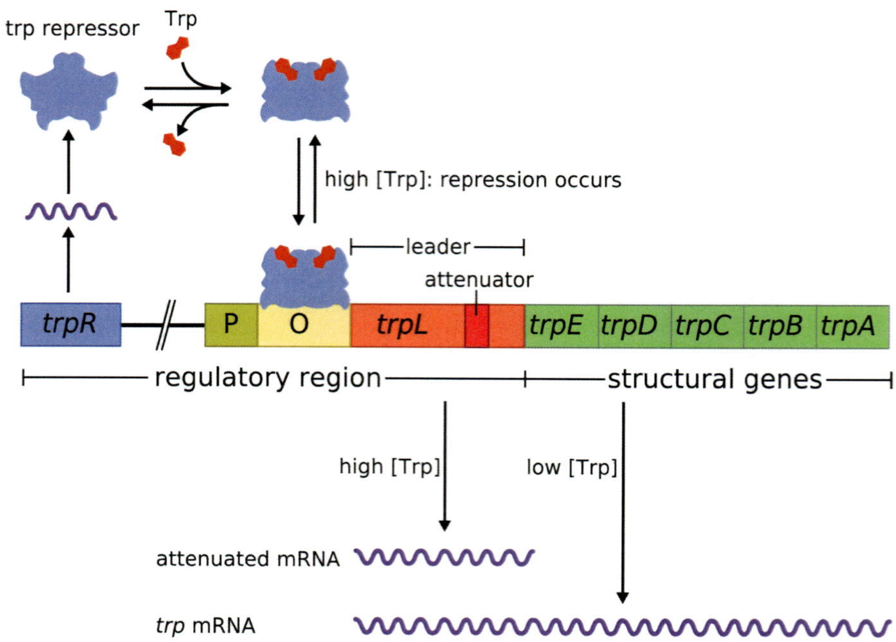

Figure 4.3. *The five genes that are needed to synthesize tryptophan in E. coli are located next to each other in the trp operon.*

Source: By Histidine – Own work, CC BY-SA 3.0, https://commons.wikimedia.org/w/index.php?curid=13443283.

The first trp gene and the promoter are separated by a DNA sequence known as the operator. The repressor protein can bind to the DNA code found in the operator. The amount of tryptophan in the cell controls the repressor protein.

Two tryptophan molecules bind to the trp repressor when tryptophan is present in the cell. The repressor takes on a new form as a result, binding to the TRPP operator. When the tryptophan-repressor complex binds to the operator, RNA polymerase is physically prevented from binding and transcription of the genes downstream. Thus, when the cell has enough tryptophan, it is prevented from making more tryptophan.

There is no tryptophan for the repressor to bind to when tryptophan is absent from the cell. The repressor does not bind to the operator or become active. RNA polymerase can therefore transcribe the operon and produce the enzymes needed to synthesize tryptophan. Because of this, the cell produces its own tryptophan when it is insufficient.

4.1.2.2. The lac Operon: An Inducer Operon

The *lac* operon in *E. coli* has more complex regulation, involving both a repressor and an activator. *E. coli* uses glucose for food but is able to use other sugars, such as lactose, when glucose concentrations are low. Three proteins are needed to break down lactose; they are encoded by the three genes of the *lac* operon.

When lactose is not present, the proteins to digest lactose are not needed. Therefore, a repressor binds to the operator and prevents RNA polymerase from transcribing the operon.

When lactose is present, lactose binds to the repressor and removes it from the operator. RNA polymerase is now free to transcribe the genes necessary to digest lactose (Figure 4.4).

However, the story is more complex than this. Since *E. coli* prefers to use glucose for food, the *lac* operon is only expressed at low levels even when the repressor is removed. But what happens when ONLY lactose is present? Now the bacterium needs to ramp up production of the lactose-digesting proteins. It does so by using an activator protein called catabolite activator protein (CAP).

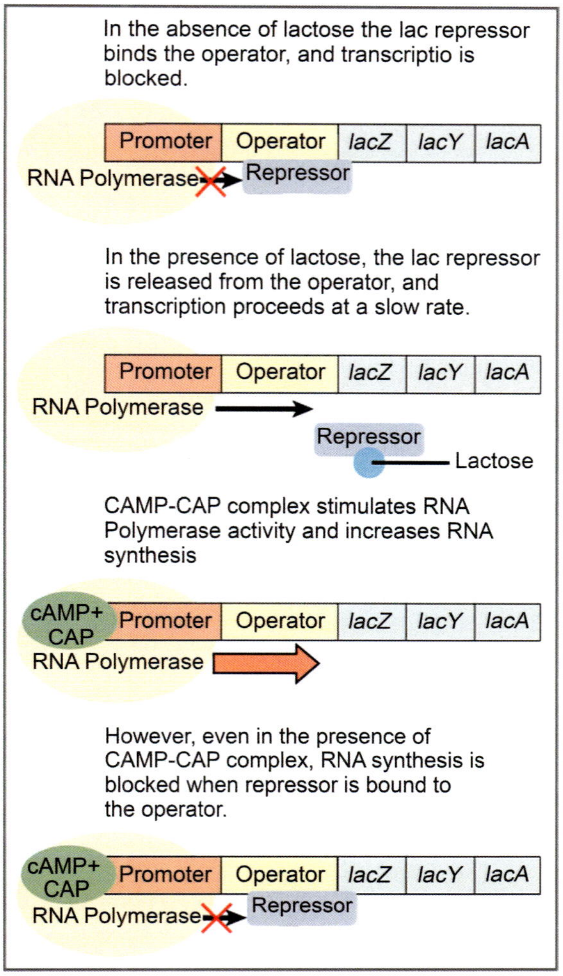

Figure 4.4. *Transcription of the lac operon only occurs when lactose is present. Lactose binds to the repressor and removes it from the operator.*

Source: https://courses.lumenlearning.com/suny-wmopen-biology1/chapter/outcome-prokaryotic-gene-regulation/.

In response to a decrease in glucose, the cell starts to accumulate cyclic AMP (cAMP). The lac operon promoter is bound by the complex formed when cAMP binds to CAP (Figure 4.5). This boosts RNA polymerase's capacity to bind to the promoter and speeds up gene transcription.

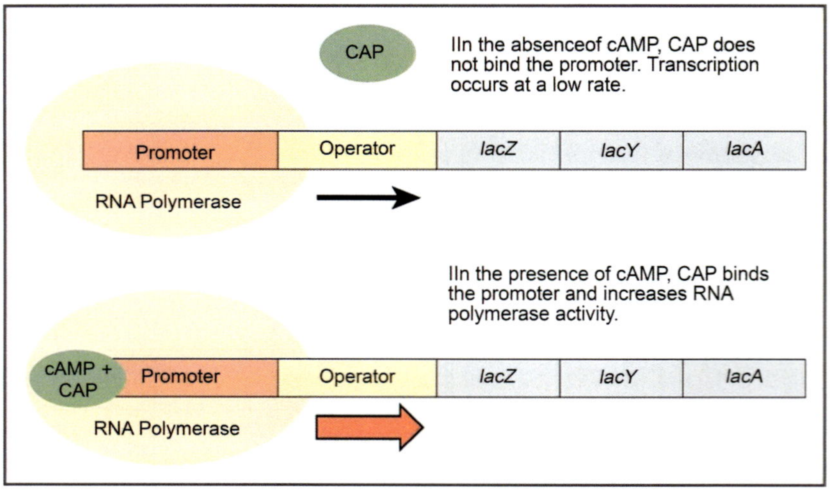

Figure 4.5. *When there is no glucose, the CAP activator increases transcription of the lac operon. However, if no lactose is present, the operon is not activated.*

Source: https://courses.lumenlearning.com/suny-wmopen-biology1/chapter/outcome-prokaryotic-gene-regulation/.

Table 4.2. Summary of Signals that Induce or Repress Transcription of the Lac Operon

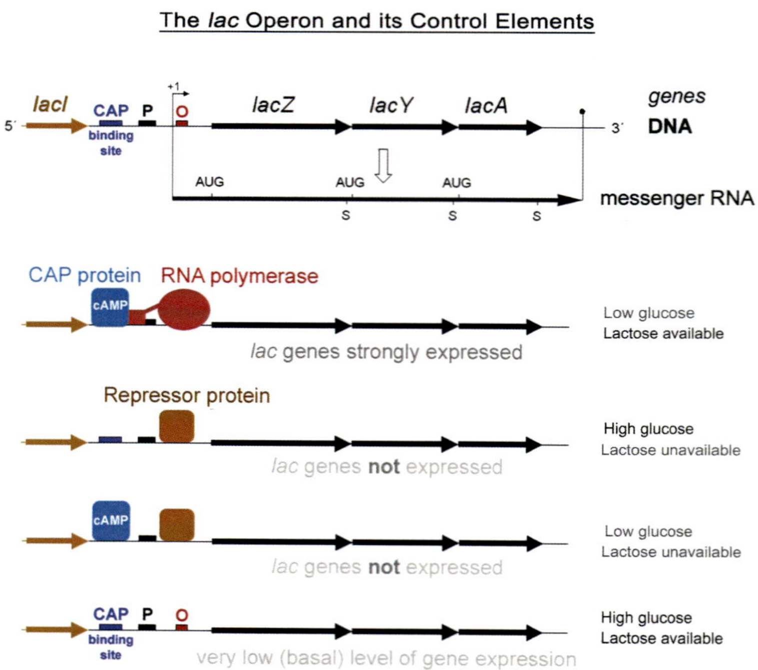

Source: By Lac_operon.png: G3proderivative work: Tereseik (talk) – Lac_operon.png, CC BY 2.0, https://commons.wikimedia.org/w/index.php?curid=9038377.

To summarize, two requirements need to be satisfied for the lac operon to become fully active. The glucose must first be either completely absent or extremely low. There must also be lactose. The maximum transcription of the lac operon occurs only in the absence of glucose and in the presence of lactose. This makes physiological sense for the cell, as producing the proteins needed to break down lactose in the event of a glutamate shortage or abundance would be energy-dense (Table 4.2).

Keyword

Catabolite Activator Protein (CAP) – It is a transcriptional activator, present as a homodimer in solution, each subunit including a ligand-binding domain at the N-terminus and a DNA-binding domain at the C-terminus.

4.1.3. Eukaryotic Gene Regulation

In higher organisms, such as eukaryotes, gene expression regulation is more intricate and dynamic. The regulation of each eukaryotic gene must occur independently since they are not arranged into operons. A lot more genes are present in eukaryotic cells than in prokaryotic ones. In the process of translating DNA into mRNA and mRNA into protein, regulation of gene expression can occur at any point. Regulation can be categorized into five levels for ease of use: transcriptional, post-transcriptional, translational, and epigenetic (Figure 4.6).

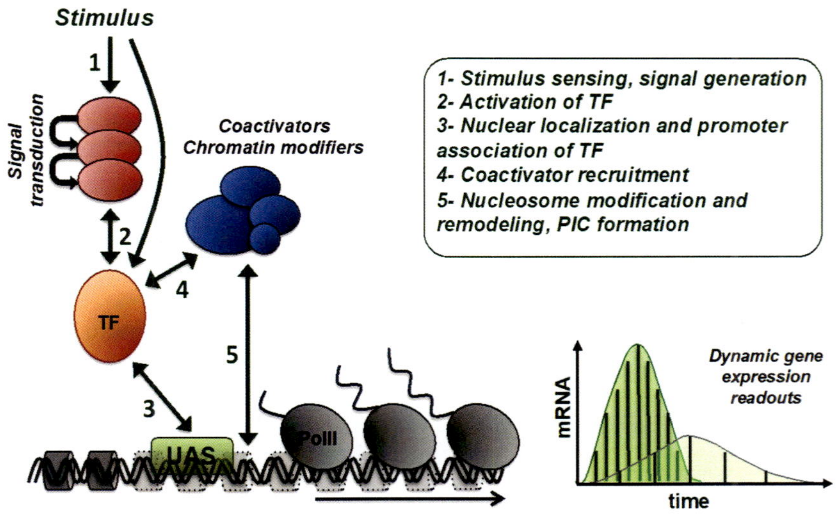

Figure 4.6. *Regulation of gene expression in eukaryotes can occur at five different levels. Here, the Central Dogma is diagrammed with arrows showing where each type of eukaryotic regulation of gene expression interrupts it.*

Source: Pascual-Ahuir, A., Fita-Torró, J., & Proft, M., (2020). Capturing and understanding the dynamics and heterogeneity of gene expression in the living cell. International Journal of Molecular Sciences, 21(21), 8278. https://doi.org/10.3390/ijms21218278.

4.1.3.1. Epigenetic Control of Gene Expression

Epigenetic regulation, or "around genetics," is the first level of control over gene expression. The field of epigenetics in biology is relatively new, but it is expanding.

Gene modifications that are temporary and do not affect the DNA's nucleotide sequence are referred to as epigenetic control. Rather, these modifications modify the chromosomal structure to enable the on/off regulation of genes. This degree of regulation is achieved by inherited chemical changes in the chromosomal proteins and/or DNA.

The addition of methyl groups to DNA, a process known as methylation, is one example of how DNA can be chemically modified. Methylation typically inhibits transcription. It is interesting to note that methylation patterns can be inherited by new cells. Therefore, a parent's propensity for a gene to express itself may be inherited by their child. It is possible for DNA to undergo additional heritable chemical changes.

4.1.3.2. Modification of Histone Proteins Is an Example of Epigenetic Control

Histone protein modification is the most researched application of epigenetic regulation. Chromosome proteins called histones cinch DNA tightly so that it can fit inside a cell's nucleus. For instance, there are more than 3 billion nucleotide pairs in the human genome. There are 46 chromosomes in each body cell, and an average chromosome has 130 million nucleotide pairs. An average human chromosome would be more than four centimeters long if it were stretched out linearly. The DNA has to be tightly wound around proteins in order to fit all of this information into the nucleus of a microscopic cell. Additionally, it is set up so that different cell types can access different segments as needed (Figure 4.7).

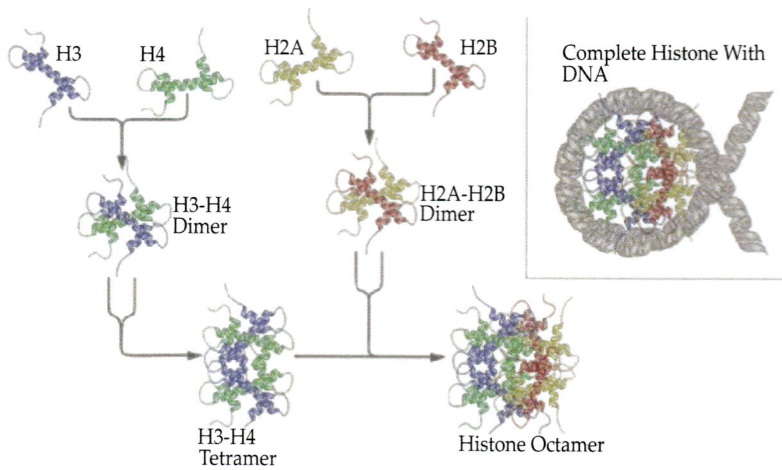

Figure 4.7. *In each chromosome, DNA is wound around histone proteins to pack it into the nucleus of a cell.*

Source: By Richard Wheeler (Zephyris) – English Wikipedia, CC BY-SA 3.0, https://commons.wikimedia.org/w/index.php?curid=1234552.

The first level of organization, or packing, is the winding of DNA strands around histone proteins. Histones package and order DNA into structural units called nucleosome complexes, which can control the access of proteins to the DNA regions (Figure 4.8). Under the electron microscope, this winding of DNA around histone proteins to form nucleosomes looks like small beads on a string (Figure 4.8). These beads (histone proteins) can move along the string (DNA) and change the structure of the molecule.

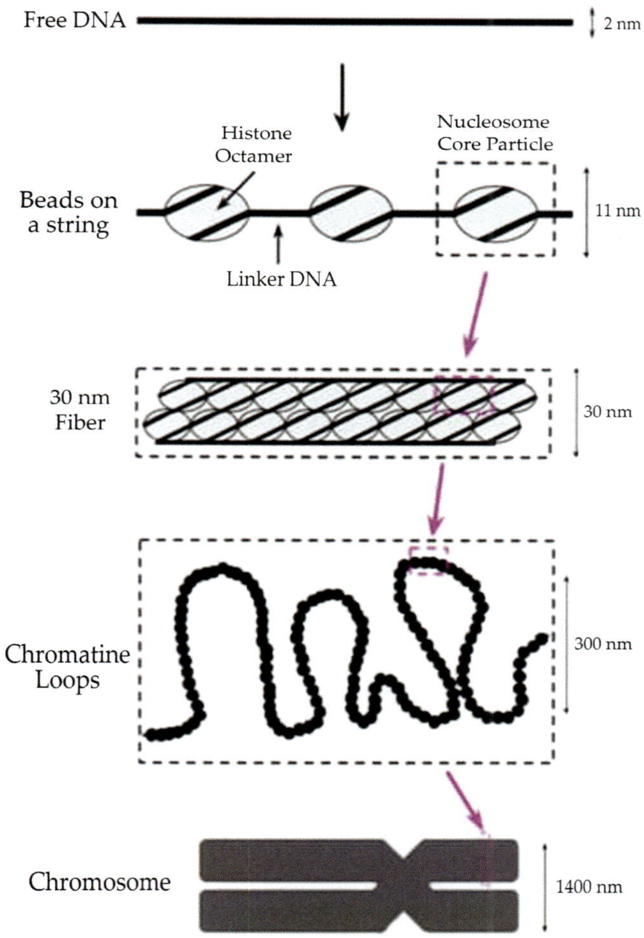

Figure 4.8. *DNA is wrapped around histones to create nucleosomes, which control the access of proteins to DNA. When viewed through an electron microscope, the nucleosomes look like beads on a string.*

Source: Morgan, A., LeGresley, S., & Fischer, C., (2021). Remodeler catalyzed nucleosome repositioning: Influence of structure and stability. International Journal of Molecular Sciences, 22(1), 76. https://doi.org/10.3390/ijms22010076.

To initiate transcription, RNA polymerase and other proteins known as transcription factors must attach to the promoter region of a gene by sliding down the DNA, which opens the chromosomal region. This process occurs when nucleosomes surrounding a gene slide down the DNA. DNA and histone proteins undergo distinct changes that indicate a closed chromosomal arrangement when a gene is to stay inactive, or silenced. When the configuration is closed, transcription cannot take place because RNA polymerase and transcription factors are unable to access the DNA (Figure 4.9).

Signals present on the histone proteins control how the proteins move. These signals are "tags" that open or close a chromosomal region; they can take the shape of phosphate, methyl, or acetyl groups (Figure 4.9). These tags are addable and removable as needed; they are not set in stone. Since DNA is negatively charged, the degree to which the DNA molecule is wound will vary depending on the charge of the histone. Histone proteins have a significant positive charge when they are unmodified; this charge is reduced when acetyl groups or other chemical modifications are added.

Keyword

Nucleosome: A nucleosome is the basic repeating subunit of chromatin packaged inside the cell's nucleus.

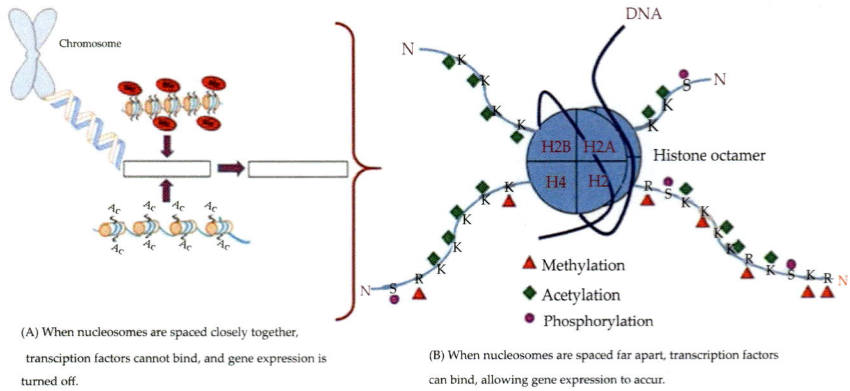

Figure 4.9. *Nucleosomes can slide along DNA.*

Note: (A) When nucleosomes are spaced closely together, transcription factors cannot bind, and gene expression is turned off. (B) When nucleosomes are spaced far apart, transcription factors can bind, allowing gene expression to occur.

Source: Rajan, P. K., Udoh, U. A., Sanabria, J. D., Banerjee, M., Smith, G., Schade, M. S., Sanabria, J., Sodhi, K., Pierre, S., Xie, Z., et al., (2020). The role of histone acetylation–methylation-mediated apoptotic gene regulation in hepatocellular carcinoma. International Journal of Molecular Sciences, 21(23), 8894. https://doi.org/10.3390/ijms21238894.

Did you get it?

1. **Discuss the Epigenetic Control of Gene Expression**
2. **How to Modification of Histone Proteins Is an Example of Epigenetic Control?**

4.2. TRANSCRIPTIONAL AND POST-TRANSCRIPTIONAL CONTROL

Learning Objectives

- Discuss the Transcriptional Control of Gene Expression
- Explain the Post-Transcriptional Control of Gene Expression

The precise timing and amount of protein production within cells are ensured by two essential processes that regulate gene expression at different stages: transcriptional regulation and post-transcriptional regulation.

The transcription phase, when RNA polymerase creates RNA molecules from DNA templates, is when transcriptional regulation takes place. The mechanism entails transcription factors attaching themselves to certain DNA sequences that are close to the genes that need to be transcribed. These sequences are referred to as enhancers and promoters. By permitting or preventing RNA polymerase from being recruited to the gene promoter region, transcription factors have the ability to either promote or suppress transcription. Gene expression can also be impacted by chromatin structure and modifications, such as histone acetylation and DNA methylation, which alter DNA's accessibility to transcription factors. Cells can react to different stimuli and signals by modifying the synthesis of particular RNA transcripts through a process known as transcriptional regulation.

Substance processing, modification, and stability are all part of post-transcriptional regulation, which happens after transcription. RNA splicing is one such process that creates mature mRNA transcripts by cutting out introns and joining exons together. Increased diversity of proteins encoded by the genome can be achieved through alternative splicing, which produces multiple mRNA isoforms from a single gene. The stability, localization, and translational efficiency of mRNA can be affected by additional post-transcriptional changes such as methylation, RNA editing, and the insertion of a poly-A tail and 5′ cap. Moreover, mRNA transcripts can bind to non-coding RNAs like microRNAs (miRNAs) and long non-coding RNAs (lncRNAs), which then control the translation or degradation of each transcript. Generally, to ensure that cells and organisms develop, function, and adapt appropriately, transcriptional and post-transcriptional regulation cooperate to fine-tune gene expression in response to internal and external stimuli. Understanding the molecular mechanisms underlying gene expression control is crucial because dysregulation of these processes can result in a number of diseases, such as cancer, neurodegenerative disorders, and metabolic syndromes.

4.2.1. Transcriptional Control of Gene Expression

The regulation of transcription is the process by which a gene in a given cell determines whether or not an mRNA is transcribed. Enzyme binding to a promoter is necessary for the initiation of transcription in eukaryotic cells, just like in prokaryotic cells. To promote transcription initiation in eukaryotes, RNA polymerase needs additional proteins known as transcription factors. In order to regulate the target gene's transcription, transcription factors are proteins that bind to the promoter and other regulatory regions. Transcription cannot be started in eukaryotic cells by RNA polymerase alone. Before transcription can start, transcription factors need to attach to the promoter region and draw RNA polymerase to the area.

4.2.1.1. The Promoter and Transcription Factors

The promoter region is located right upstream of the coding sequence in eukaryotic genes. A few to hundreds of nucleotides can make up this region. Gene-specific variations can be observed in the length of the promoter. More space is available for proteins to bind to the longer the promoter. As a result, there can be significant differences in the degree of control over gene expression between genes. The promoter binds transcription factors that regulate the start of transcription (Figure 4.10).

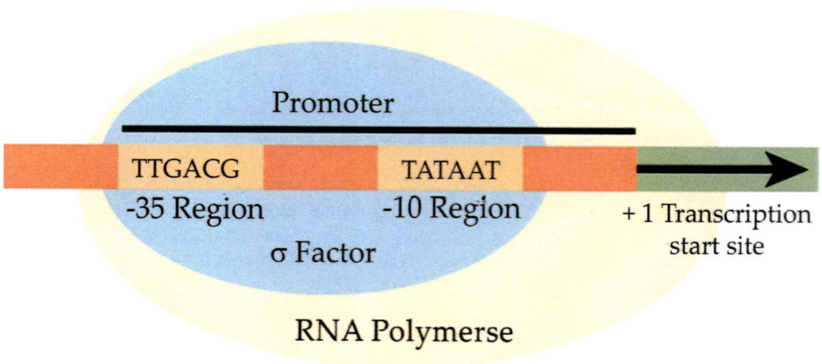

Figure 4.10. *Each gene has a promoter upstream of the coding sequence. The promoter binds to transcription factors and helps RNA polymerase to bind and start transcription. Many genes also have upstream enhancers. Enhancers bind activators, bend around, and help RNA polymerase start transcription.*

Source: https://courses.lumenlearning.com/wm-biology1/chapter/prokaryotic-transcription/.

The TATA box is located within the promoter region, directly upstream of the transcriptional start site. TATA repeats, or thymine and adenine dinucleotides, are all that are contained in this box. An initiation complex is put together by transcription factors binding to the TATA box. RNA polymerase binds to its upstream sequence and gets phosphorylated once this complex is put together. This causes the transcription initiation complex to become active, releases a portion of the protein from the DNA, and positions RNA polymerase in the proper orientation to start transcription (Figure 4.10).

4.2.1.2. Enhancers and Repressors

Certain sections aid in boosting transcription in certain eukaryotic genes. Even thousands of nucleotides apart, these areas—known as enhancers—may be found in close proximity to the genes. A gene may have them upstream, downstream, or inside the coding region. A binding site for activators is an enhancer. Enhancers that are distant from genes can interact with the transcription initiation complex through DNA folding that brings the enhancer and promoter closer (Figure 4.10).

Prokaryotic and eukaryotic cells both possess transcription-inhibiting mechanisms. Transcription can be inhibited by transcriptional repressors by binding to enhancer or promoter regions. Which genes need to be expressed depends on how both activators and repressors react to outside stimuli.

Did you know?

Although as early as 1951, Barbara McClintock showed interaction between two genetic loci, Activator (Ac) and Dissociator (Ds), in the color formation of maize seeds, the first discovery of a gene regulation system is widely considered to be the identification in 1961 of the lac operons, discovered by François Jacob and Jacques Monod, in which some enzymes involved in lactose metabolism are expressed by E. coli only in the presence of lactose and absence of glucose.

4.2.2. Post-Transcriptional Control of Gene Expression

After the mRNA is transcribed but before translation starts, there is post-transcriptional regulation. This control may take place at the level of ribosome binding, mRNA processing, or transport from the nucleus to the cytoplasm.

4.2.2.1. Alternative RNA Splicing

A process known as alternative RNA splicing makes it possible to cut different combinations of introns and occasionally exons from the main transcript (Figure 4.11). This enables the production of multiple protein products from a single gene. Gene regulation may be accomplished through alternative splicing.

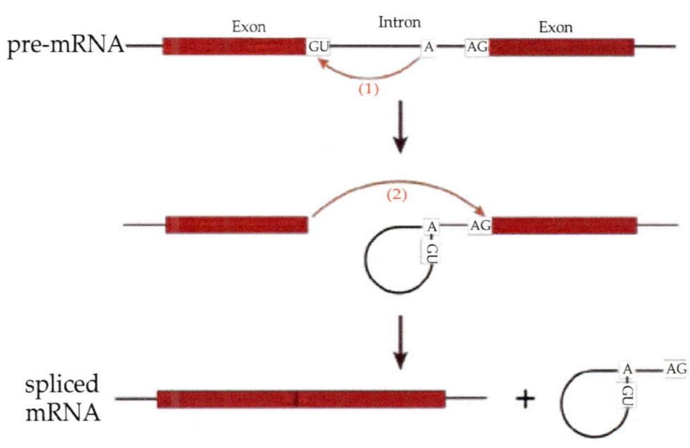

Figure 4.11. *Before RNA can be translated, introns must be removed by splicing. Pre-mRNA can be alternatively spliced to create different proteins.*

Source: By BCSteve – Own work, CC BY-SA 3.0, https://commons.wikimedia.org/w/index.php?curid=30096313.

Several protein products can be produced in various cells or at various times within a single cell by means of differential splicing. Up to 70% of human genes are expressed as multiple proteins as a result of alternative splicing, which is now recognized as a common mechanism of gene regulation in eukaryotes.

4.2.2.2. Control of RNA Stability

The stability of the mRNA in the cytoplasm is a form of post-transcriptional control. An mRNA must take longer to translate and produce more protein the longer it is present in the cytoplasm. The length of the poly-A tail of mRNA is one of several factors that affect its stability. RNA-binding proteins (RBPs) are proteins that can bind to RNA segments that are either upstream or downstream of the protein-coding region. Untranslated regions, or UTRs, are the parts of the RNA that are not translated into proteins. The 5′ UTR is the region that comes directly before the protein-coding region, and the 3′ UTR is the region that comes after the coding region (Figure 4.12). Depending on which particular RBP binds to these locations, the stability of an RNA molecule may increase or decrease as a result of the binding.

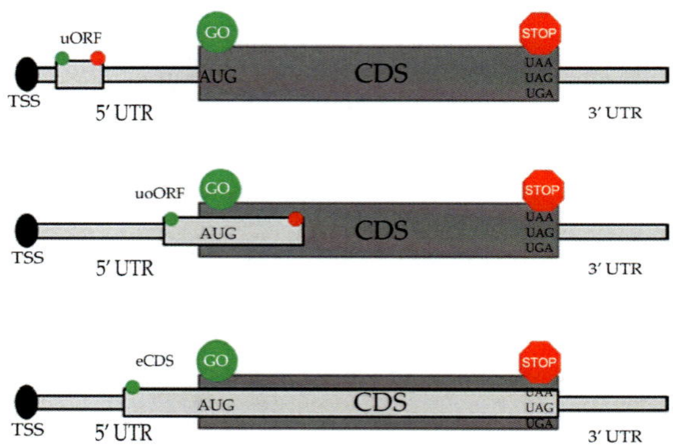

Figure 4.12. *The protein-coding region of mRNA is flanked by 5′ and 3′ untranslated regions (UTRs). RNA-binding proteins at the 5′ or 3′ UTR influence the stability of the RNA molecule.*

Source: Soukarieh, O., Meguerditchian, C., Proust, C., Aïssi, D., Eyries, M., Goyenvalle, A., &Trégouët, D. A., (2022). Common and rare 5′UTR variants altering upstream open reading frames in cardiovascular genomics. Front. Cardiovasc. Med., 9, 841032. doi: 10.3389/fcvm.2022.841032.

RNA molecules can also be bound by miRNAs or microRNAs. Dicer is a protein that cuts longer pre-miRNAs in the nucleus into shorter mature miRNAs. miRNAs are short (21–24 nucleotides) RNA molecules. The RNA-induced silencing complex (RISC), a ribonucleoprotein complex, and miRNAs bind to mRNA together. The target mRNA is quickly degraded by the RISC-miRNA complex.

Did you get it?

1. **Focus on The Promoter and Transcription Factors**
2. **Explain the Control of RNA Stability**

4.3. TRANSLATIONAL AND POST-TRANSLATIONAL

Learning Objectives

- Focus on translational control of gene expression
- Discuss the post-translational control of gene expression

Important processes that follow mRNA transcription and translation into protein include translational and post-translational regulation, which guarantee appropriate protein activity regulation and proper cellular functioning.

Translational regulation takes place during the synthesis of proteins when the mRNA's nucleotide sequence is decoded to put together a particular arrangement of amino acids to form a polypeptide chain. Numerous factors, such as the presence of regulatory elements in the mRNA sequence, the activity of ribosomes, and the availability of mRNA transcripts, regulate this process. As the first stage of protein synthesis, translational initiation is a crucial regulatory event that is managed by ribosome-binding proteins and initiation factors. Cells can adapt their protein synthesis to suit their own requirements, those of the developing embryo, and changes in the environment thanks to translational regulation.

Post-translational regulation pertains to the alteration and handling of proteins after their synthesis. This covers a broad variety of covalent modifications, including glycosylation, ubiquitination, acetylation, phosphorylation, and proteolytic cleavage. The alterations in protein structure, stability, localization, and activity can impact the way proteins function and interact with other components of the cell. For instance, while ubiquitination directs proteins toward the proteasome for degradation, phosphorylation of proteins by kinases can control enzymatic activity, signaling pathways, and protein-protein interactions. Post-translational modifications are essential for many different biological processes, including metabolism, gene expression, cell signaling, and many others.

In general, cellular homeostasis, responding to changing environmental stimuli, and fine-tuning gene expression all depend on translational and post-translational regulation. These processes are crucial to cellular physiology and disease pathology because their dysregulation can result in a number of illnesses, such as cancer, neurodegenerative diseases, and metabolic syndromes. Comprehending the mechanisms that underpin translational and post-translational regulation offers valuable perspectives into the intricacy of cellular operations and presents possible avenues for therapeutic interventions.

Keyword

Cell Signaling – It is the process by which a cell interacts with itself, other cells, and the environment.

4.3.1. Translational Control of Gene Expression

Translated into proteins, mRNAs are produced once they reach the cytoplasm. mRNA plays a major role in controlling this process. As we've already covered, the translation of mRNA into a protein is significantly influenced by its stability. At the level where the mRNA binds to the ribosome, translation can also be controlled. Both the rate and degree of translation can be altered after the mRNA has attached itself to the ribosome. Proteins intended for the endoplasmic reticulum (ER), an organelle, are an example of translational control in action. A signal sequence is a tag consisting of the initial few amino acids of these proteins. Following the translation of these amino acids, the mRNA-ribosome complex is transported to the end of the replication machinery (ER) and a signal recognition particle (SRP) attaches to the signal sequence to halt translation. The translation proceeds after they arrive, removing the SRP.

Keyword

Endoplasmic Reticulum (ER) – It is a large, dynamic structure that serves many roles in the cell including calcium storage, protein synthesis, and lipid metabolism.

4.3.2. Post-Translational Control of Gene Expression

Post-translational regulation is the last stage of gene expression regulation in eukaryotes. This kind of control entails altering the protein after it is produced in order to change its activity. Enzyme inhibition is one type of post-translational regulation. A competitive or allosteric inhibitor inhibits an enzyme when it is no longer required by preventing it from binding to its substrate. The enzyme can be reactivated later on because the inhibition is reversible. Compared to breaking down the enzyme when not needed and then producing more when needed, this is more efficient.

Functional groups like methyl, phosphate, or acetyl groups can also be added to proteins to control their activity and/or stability. Occasionally, these alterations can control a protein's location within the cell, such as its attachment to the plasma membrane, presence in the cytoplasm, or nucleus.

When a protein acquires a ubiquitin group, it is designated for protein degradation. Like a flag, ubiquitin signifies that the protein's life cycle is coming to an end. Proteases are an organelle that breaks down proteins; tagged proteins are transferred to them (Figure 4.13). Therefore, changing the protein's half-life is one method of regulating gene expression.

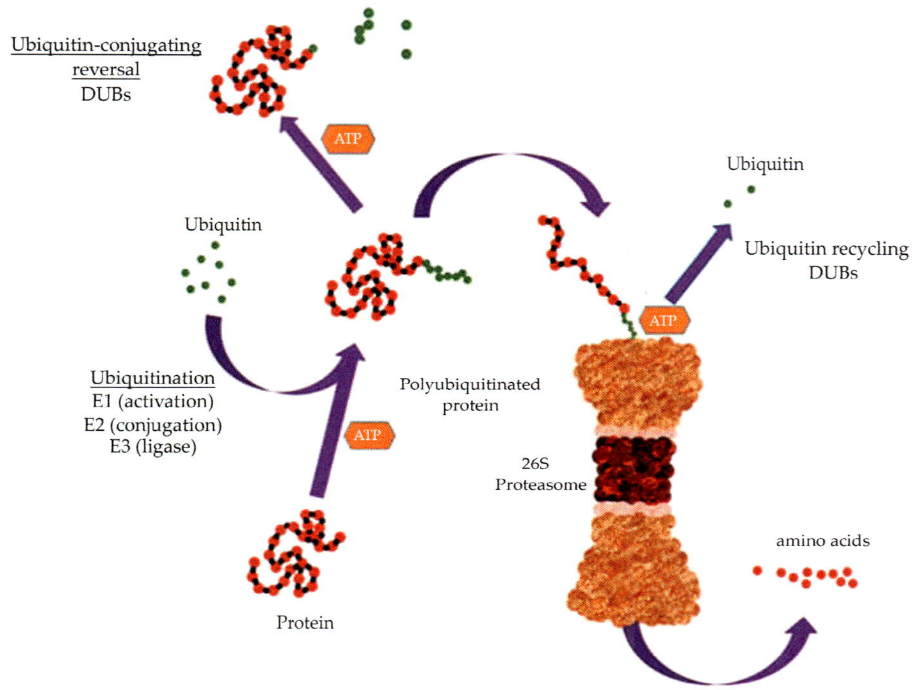

Figure 4.13. *Proteins with ubiquitin tags are marked for degradation within the proteasome.*

Source: Fhu, C. W., & Ali, A., (2021). Dysregulation of the ubiquitin proteasome system in human malignancies: A window for therapeutic intervention. Cancers, 13(7), 1513. https://doi.org/10.3390/cancers13071513.

4.4. EPIGENETICS: MECHANISMS AND IMPLICATIONS

Learning Objectives

- Focus on mechanisms;
- What do you mean by drosophila?

Epigenetics is the study of heritable modifications to cellular phenotype or gene expression without corresponding changes to the underlying DNA sequence. Modulations of non-coding RNAs, which together affect chromatin structure and gene activity, as well as chemical alterations to DNA and histone proteins, are responsible for these modifications. Gene imprinting, cellular differentiation, development, and disease pathogenesis are just a few of the biological processes in which epigenetic modifications are crucial.

DNA methylation, which is the addition of a methyl group to cytosine bases, particularly in the context of CpG dinucleotides, is one of the most extensively researched epigenetic mechanisms. DNA methyltransferase enzymes create and maintain DNA methylation patterns, which are essential regulators of gene expression. Gene promoter hypermethylation is frequently linked to transcriptional silencing, whereas hypomethylation can result in gene activation. Deviant DNA methylation patterns have been linked to many illnesses, such as cancer, neurological conditions, and abnormalities in development.

Histone modifications, which include the covalent attachment of chemical groups like acetyl, methyl, and phosphate groups to histone proteins within chromatin, are another important epigenetic mechanism. These changes affect how accessible DNA is to transcription factors and other transcriptional machinery, which in turn affects how genes are expressed. Histone methylation, for instance, can have both activating and repressive effects, depending on the precise lysine or arginine residues targeted, whereas histone acetylation is typically linked to active gene transcription.

Moreover, non-coding RNAs—such as long non-coding RNAs (lncRNAs) and microRNAs—modify gene expression at the post-transcriptional level, which is how they contribute significantly to epigenetic regulation. While long noncoding RNAs (lncRNAs) can control chromatin structure, transcriptional activity, and mRNA processing through a variety of mechanisms, miRNAs can bind to target mRNAs and inhibit translation or promote degradation. Non-coding RNAs are important for maintaining cellular homeostasis and epigenetic regulation, as evidenced by the multiple diseases linked to their dysregulated expression. The field of epigenetics comprises an intricate web of molecular processes that regulate gene expression and cellular characteristics in addition to the DNA sequence. Personalized medicine and the creation of innovative therapeutic approaches stand to benefit greatly from our growing understanding of the roles played by epigenetic modifications in health and disease.

Remember

Histone modifications are essential for the regulation of genes and the proper operation of cells; their dysregulation has been linked to a number of diseases.

The term "epigenetics," first used by Conrad Waddington in 1942 (Figure 4.14), was later defined as the causal relationships between genes and their products that allow for the expression of certain phenotypes.

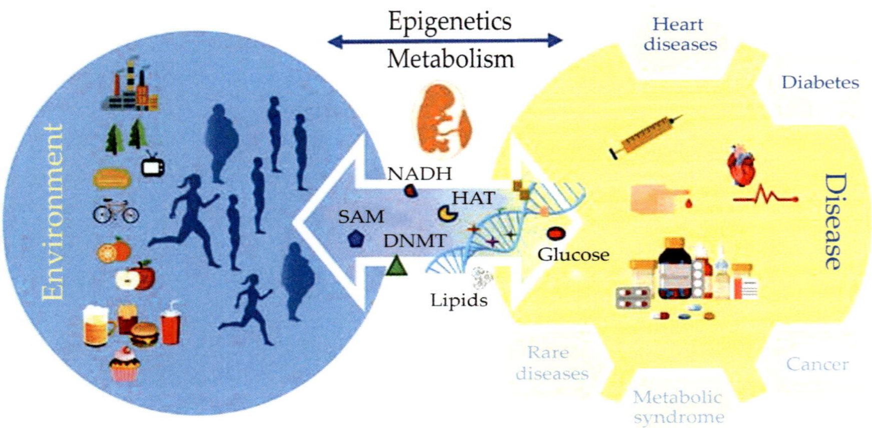

Figure 4.14. *Epigenetics and health.*

Source: Tzika, E., Dreker, T., & Imhof, A., (2018). Epigenetics and metabolism in health and disease. Front. Genet., 9, 361. doi: 10.3389/fgene.2018.00361.

While the term "epigenetics" has many different definitions these days, the accepted definition states that it encompasses all heritable phenotypic changes caused by mechanisms unrelated to the core DNA sequence. It was long thought that the only way epigenetic information could be passed down was during cellular divisions. But it is now established that organisms are able to transmit epigenetic processes down to their progeny. It is the DNA methyltransferases (DNMTs) that catalyze this important transfer step. DNMT1 catalyzes the methylation that occurs as the DNA is copied during each cellular replication. DNMT1, 3A, and 3B are the three main DNMTs. Since de novo methylation, which happens when new 5-methyl cytosines are inserted into the genome at sites that have not yet been methylated, is more well-known, the other two main DNMTs, 3A and 3B, are more well-known for it. More methylation in a gene regulatory region increases the likelihood that the gene activity will be down-regulated, and vice versa, with a few notable exceptions to this generalization. A combination of 5-hydroxymethylcytosine and 5-methylcytosine can express the effects of DNA methylation on the genome. Recent studies have highlighted the critical roles of UHRF1 and DNMT3L, which are required for maintenance and de novo methylation, respectively.

4.4.1. Mechanisms

Chromatin modifications are another crucial epigenetic mechanism that impacts gene expression along with numerous other biological functions. Post-translational modifications to histones, such as acetylation and methylation, are site-specific and impact the activities of other proteins involved in gene regulation. Histone acetyltransferases (HATs) acetylate key histones that comprise chromatin, while histone deacetylases (HDACs) deacetylate them. Although these changes can happen anywhere in the histones, they are most commonly observed in the protein's amino-terminal regions. Gene activity and histone acetylation are often correlated, and vice versa. For example, non-coding RNA, in both its long and short forms, often shares RNA and protein components with the RNA interference (RNAi) pathway and may also affect DNA methylation and other aspects of epigenetics. From protists to humans, a broad range of organisms appear to be impacted by these impacts. The epigenetic effects of prions are intriguing because they are not dependent on chromatin or DNA.

Since altered states of regulatory networks or macromolecular complexes can influence phenotypic expression without altering DNA sequences, structural heredity is essential for epigenetic expression. Epigenetic expression can be impacted and human disease can result from the prion proteins' autocatalytic process of structural modification. Additionally, a gene's position on a specific chromosome can affect how that gene expresses itself.

4.4.1.1. Advancements

Significant progress has also been made in high-resolution methylome analysis. With the use of methylation-sensitive restriction enzymes, sodium bisulfite conversion, and affinity capture with antibodies or proteins that select methylated DNA sequences, new techniques for analyzing genome-wide aspects of epigenetics have been made possible by high-throughput sequencing and microarray platforms. Methylated DNA immunoprecipitation (MeDIP), methylation-sensitive restriction fingerprinting, targeted and whole genome bisulfite sequencing, and methylation-specific digital karyotyping were developed.

The analysis of the spatial proximity of distant functional genomic locations can create a three-dimensional picture of the genome inside the nucleus. Because so much information is being gathered from epigenomic technologies, techniques for interpreting data from ChIP-on-chip and ChIP-seq, for instance, are gaining importance. The development of epigenetics has undoubtedly been and will continue to be significantly influenced by advancements in the techniques for analyzing epigenetic data. The study of epigenetics is based primarily on model systems that can be both typical and exceptional, and epigenetic processes are widely observed. The discovery of the molecular mechanisms behind fundamental epigenetic events has been facilitated by the ease of working with eukaryotic microorganisms. Here, epigenetics may play a key role in safeguarding the genome from invasive DNA elements and helping to understand the importance of gene silencing mechanisms in evolution.

4.4.2. Drosophila

Drosophila aids in the comprehension of the evolutionary consequences of epigenetics. For epigenetics, the mouse model is the most useful model system. Numerous mouse models, which are essential for epigenetic processes like imprinting and trans-generational epigenetics and may shed light on human diseases like diabetes, cancer, and neurological disorders, were examined by Randy Jirtle and colleagues.

Because of their flexibility and ability to mute transposable elements, plant models are essential in the field of epigenetics. Plant RNAi silencing has been a pioneer in the field of epigenetics. It is impossible to fully understand epigenetics and changes in an organism's metabolism separately because of their close relationship. Crucial to epigenetic mechanisms is the universal methyl donor S-adenosylmethionine (SAM), which drives many epigenetic processes. Metabolic activities can influence chromatin, a crucial modulator of epigenetic events. The enzymes that animals use to modify their histones are now understood to be under the influence of a wide range of environmental variables and metabolic substances.

Since metabolic processes affect DNA methylation and chromatin remodeling, the two primary epigenetic mediators, it is expected that this technology will continue to grow exponentially. Partially reliant on environmental cues, stem cells employ epigenetic mechanisms such as DNA methylation, histone modification, and microRNA (miRNA) to regulate their response to these cues. Regenerative medicine requires stem cells, and the regeneration of skeletal muscle requires substantial epigenetic modifications that regulate the expression of specific genes in muscle progenitors via chromatin and microRNA epigenetic modifications.

Comparable to genomic imprinting, epigenetics has long been known to be involved in this process, though the field's research is still developing swiftly. Genomic imprinting can happen in particular tissues or during critical developmental stages. Similar processes also cause it to occur in plants and other animals besides mammals. Recently, there has been progress in the field of memory processing epigenetics. Interesting new findings have shown that epigenetic processes regulate genes to produce changes in adult brain function and behavior based on life events. Furthermore, cutting-edge drugs that modify epigenetic pathways might be employed in the future to treat or mitigate cognitive impairment.

Since early life events can have a significant impact on epigenetic markers and adult health and disease risk, transgenerational inheritance is also a sort of memory based in part on epigenetics. Because our parents' genes have an epigenetically influenced effect on our genes throughout our lives, the aging process is a form of epigenetic memory and experience. This may affect how long people live and how likely we are to get a number of age-related illnesses, many of which are also epigenetically based. Consequently, it is evident that epigenetics influences a wide range of unique activities, and it is likely that many more will be found in the future. The impact of epigenetics on evolution is still great today because of the dynamic environment that can affect gene expression through epigenetic processes. Fast dietary changes, contemporary lifestyles, and environmental pollution, for example, affect not only the epigenome of humans but also the evolution of many more rudimentary species, which have a major effect on the environment. There are significant differences in dietary components not only between individuals but also within human communities.

The epigenetic expression of the genome, for example, is influenced by bacteria and viruses, and these modifications can lead to diseases such as cancer in humans. The details of how adenovirus type 12 (Ad12) changed the hamster genome are discussed by Walter Doerfler and associates. Because medications also change the epigenome, a new field called pharmacoepigenetics has emerged. It is clear that different populations respond to medications in different ways, and epigenetic factors may account for a significant portion of this variation. Medical professionals' interest in epigenetics has grown as a result of its potential implications for understanding the genesis, course, and diagnosis of human diseases. Long associated with epigenetic modifications, cancer has been shown to be influenced by DNA methylation, chromatin modifications, and RNA-dependent regulation. Numerous autoimmune diseases, such as multiple sclerosis, and immunological disorders, such as rheumatoid arthritis and systemic lupus erythematosus (SLE), have been associated with epigenetic abnormalities. Brain disorders have also been linked to epigenetic processes. Numerous illnesses, including Rett syndrome, Alzheimer's disease, Huntington's disease, and even autism, have been related to epigenetic modifications.

Remember

Epigenetics may even play a role in the expression of depression and schizophrenia. System metabolic abnormalities may have a connection to epigenetic abnormalities problems.

For example, fetal chromatin may be impacted by hypertension, gestational diabetes, and obesity, which could increase the risk of adult diseases later in life. Given that epigenetic mechanisms underlie genomic imprinting, it is not surprising that errors in imprinting could lead to a range of human disorders. Epigenetic abnormalities can result in imprinting disorders such as Prader-Willi syndrome, Angelman syndrome, silver Russell syndrome, and many others, including transient neonatal diabetes mellitus. Because epigenetic factors influence so many diseases, advancements in the field of epigenetic therapy will likely play a part in improving their management.

4.4.3. Drugs

While other epigenetic therapies are being used, the most attention in the clinics has probably been focused on histone-modifying medications. The most significant of them are inhibitors of histone deacetylase (HDAC). There are many different types of HDAC inhibitors available, and it is likely that significant progress will be made in treating abnormalities in DNA methylation, histone modifications, and possibly some of the numerous other epigenetic processes that have been discovered, in addition to HDAC inhibitors. Methyltransferases are among the protein-modifying enzymes that alter histones, particularly at their flexible N-termini. When it comes to distributed retrotransposons and altered sex-specific de novo methylation of DNA sequences in germ cells, the phenotype of Dnmt3L knockout mice is identical to that of a Dnmt3a germ cell-specific conditional knockout. According to these findings, Dnmt3a and Dnmt3L are both necessary for the methylation of the majority of imprinted loci in germ cells. Dnmt3L promotes de novo methylation by both of these MTases and co-localizes and immunoprecipitates with Dnmt3a and Dnmt3b. The C-terminal domains of Dnmt3a and Dnmt3L contain the distinctive fold of Class I AdoMet-dependent MTases. However, only Dnmt3a and not Dnmt3L contained S-adenosyl-L-homocysteine, a byproduct of the methylation reaction. This is consistent with Dnmt3a serving as the complex's catalytic component and Dnmt3L being dormant and unable to bind AdoMet. The total length of the Dnmt3a/Dnmt3L C-terminal complex is approximately 16 nm, longer than a core nucleosome, which has a diameter of 11 nm. Two Dnmt3a monomers and two Dnmt3L monomers are present in this complex, generating a tetramer with two 3L-3a interfaces and one 3a-3a interface.

Dnmt3a and Dnmt3b have the shortest DNA binding domains of all known active DNA MTases. But when two active sites are brought together by dimerization at the 3a-3a contact, the DNA-binding surface is virtually doubled. The frequency of CpG sites in the differentially methylated regions of 12 maternally imprinted mouse genes has a similar pattern. These findings point to a rationale for the detection of both CpG spacing and nucleosome alteration in order to identify and methylate differentially methylated sites in imprinted genes. When embryonic stem cells were forced to differentiate, non-CG methylation disappeared, but it reappeared in induced pluripotent stem cells. Similar to this, it has been discovered that Arabidopsis thaliana DRM2, which is related to mammalian Dnmt3a, exhibits a 10-bp correlation of non-CpG DNA methylation. Histone modifications and DNA methylation are closely related to one another. DNA methylation is actually thought to be linked with histone methylation. Predominantly, the absence of H3K4 methylation and the presence of H3K9 methylation are associated with DNA methylation. Gene promoters may be shielded from de novo DNA methylation in somatic cells by histone H3 lysine 4 (H3K4) methylation.

The H3K4 methylation patterns, which are not known to be preserved during chromosome replication, could be converted into heritable patterns of DNA methylation that mediate transcriptional silencing of the affected sequences by the mammalian de novo DNA methylation Dnmt3L-Dnmt3a machinery. *In vitro*, Dnmt3a is completely active on nucleosomal DNA. In addition to being necessary for genomic imprinting, Dnmt3a2 is a shorter variant of Dnmt3a that is predominate in embryonic stem cells and embryonal carcinoma cells and is also found in the testis, ovary, thymus, and spleen. The four core histones Dnmt3a2 and Dnmt3b were found to be the primary *in vivo* interaction partners of epitope-tagged Dnmt3L. When the amino tail of H3 and Dnmt3L were crystallized together, it was clear that the H3 amino tail had bound to the Cys-rich 3-Zn binding domain of Dnmt3L. Substituting specific residues in the binding site also rendered the H3-Dnmt3L association ineffective.

4.4.4. Enzymes

As a conserved technique, base flipping is widely used by nucleotide modifying enzymes, including DNA MTases, DNA repair enzymes, and RNA modification enzymes. No structural or sequence similarities seem to exist between the DNA MTase domain and the SRA (or of DNA repair enzymes). Nonetheless, there are similarities between DNA MTases and the flipped base binding in a concave pocket, the use of two loops to approach DNA from both major and minor grooves at the same time, and the phosphodiester backbone resulting from numerous protein-phosphate bonds surrounding the flipped nucleotide. This might be highly important since the recognition sequence for the SRA domain only consists of two base pairs.

Keyword

Nucleotides – Thes are organic molecules composed of a nitrogenous base, a pentose sugar, and a phosphate.

TET1 was overproduced in human cells, which resulted in the formation of 5hmC. The level of DNA 5mC simultaneously decreased, indicating that 5hmC is an oxidation byproduct of 5mC. 5hmC was found in ES cells, and RNAi-mediated reduction of TET1 lowered this percentage. The unexpected discovery of a 5mC oxidation pathway presents a number of issues, such as whether 5mC oxidation is a significant epigenetic change, either as a final result or as a step in active DNA demethylation, as suggested by the existence of 5hmC DNA excision repair glycosylase. Targeted deletion of the Dnmt2 gene revealed that Dnmt2 is not necessary for overall DNA methylation or maintenance in embryonic stem cells. While the Dnmt2 protein expressed by bacteria failed to methylate DNA *in vitro*, the Dnmt2 protein expressed by baculoviruses either showed no detectable activity or residual action towards DNA. A DNA methyltransferase-like catalytic mechanism is used by human Dnmt2 to methylate RNA cytosine-C5, which is not surprising given the high levels of sequence and structural conservation as well as the similar chemistry of DNA and RNA cytosine. It is impossible to create a chronological timeline of events or establish causal connections in gene silencing without an understanding of the interconnections and spatial links between their modular domains, as well as whether or not inter-domain interactions contribute to target specificity.

Gene or region-specific DNA methylation analysis is necessary in a number of circumstances, and numerous methods are available. It is crucial to comprehend the particulars of each method, including the amount of DNA required, the degree of flexibility in selecting CpG sites for analysis, the potential for quantitative analysis, the degree of technical difficulty, and the associated costs. For example, to investigate DNA methylation as a potential factor in gene silencing, one should look at a particular region that controls gene expression, and one should use a technique that enables flexible region selection. If the procedure is to be used for diagnostics, it must be extremely accurate. To determine DNA methylation and separate 5-methylcytosine (Cm) from cytosine (C), a few ideas can be applied. A methyl group on a cytosine at a CpG site, or sites within restriction sites, affects the activity of restriction enzymes, which are reliant on methylation. This is the first principle. This process turns unmethylated C into uracil (U) relatively quickly, in contrast to how slowly it changes methylated C.

Remember

This technique was frequently used, but not as much in recent times as it was before bisulfite conversion-based approaches gained popularity.

For DNA methylation analysis using Southern-blot hybridization, DNA is first digested by a methylation-sensitive restriction enzyme and then hybridized with a probe specific to a specific genomic region. One can ascertain the methylation status of a restriction recognition site by monitoring the band locations of the DNA fragments surrounding the restriction sites. Due to its ability to examine multiple related sequences across the genome with a single probe, the Southern blot methodology is especially useful for analyzing repeated sequences. However, this approach only looks at a small percentage of CpG sites that are located within restriction recognition sites, utilizing a large amount of high-quality DNA in the process.

4.4.4.1. PCR

Primers positioned in genomic regions lacking CpG sites are used in PCR to amplify bisulfite-converted DNA. The converted DNA exhibits methylated and unmethylated C, respectively, at a CpG site that has undergone thymine (T) and methylation (C). This approach allows us to investigate the methylation status of every single CpG site situated in between the primers as well as the methylation patterns of numerous CpG sites contained in a single DNA molecule. This method makes it possible to analyze DNA methylation practically. The labor-intensive nature of this method and the requirement for at least 10 clones to be sequenced from each sample could be a disadvantage. It is possible for a restriction enzyme recognition site to form following bisulfite conversion. Using primers specific to methylated or unmethylated sequences, this method probes the methylation statuses of multiple CpGs at primer locations by performing PCR and determining whether a PCR product is present or absent. Even in the case of perfect PCR conditions, PCR cycles ought to be restricted so long as sufficient PCR product is obtained for cloning. Excessive PCR cycles lead to denaturation of the PCR product, amplification of chimera products, and even PCR products that weren't in the template DNA when Taq polymerase activity is absent. Overuse of PCR cycles further magnifies the discrepancy in PCR efficiency between methylation and unmethylated DNA. With real-time MSP, DNA methylation levels can be precisely, sensitively, and quantitatively measured. A genomic region should be carefully selected, just like in other investigations. In the same region, primers specific to methylation or unmethylated DNA should be created. A primer's 3′ end should be at a polymorphic C/T site, and several CpG sites should be close to the 3′ end. It can be challenging to create primers that are specific to unmethylated DNA, and using the bottom strand of the other DNA strand is typically useful.

The ideal annealing temperature and magnesium concentration should be determined using the fully methylated and totally unmethylated DNA controls. When methylation-specific primers are present, there is adequate amplification of fully methylated DNA and no amplification of completely unmethylated DNA. The ideal conditions for real-time MSP can be estimated using the amplification curve and melting curve. On the melting curve, the ideal PCR conditions result in a single, sharp peak. To measure DNA methylation levels by real-time MSP, standard DNA with known quantities of DNA molecules is necessary. The first step is to eliminate any remaining primers and nucleotides from the PCR result using a gel-filtration column. Second, after the MSP PCR product has been cloned into the plasmid, a restriction enzyme linearizes the plasmid. Because the molecular weight of the PCR product or the plasmid containing the insert can be ascertained, the number of DNA molecules in a solution can be calculated. Although the process is labor-intensive, cloning a PCR product to create standard DNA has the advantages of accuracy and large quantities of available standard DNA. MSP and real-time MSP can achieve high sensitivity, i.e., detect one methylated DNA molecule out of 1000. However, there is a significant decrease in the amount of DNA molecules that can serve as PCR templates during bisulfite-mediated conversion. To be more precise, the amount of template DNA molecules found by quantitative PCR decreases to 5% to 10% of the DNA before the treatment, despite the fact that the weight of the DNA only slightly decreases. This means that when calculating the number of template DNA copies in a PCR solution, caution must be used.

Both basic and diagnostic research uses regional DNA methylation analysis. It is essential to select a suitable methodology and conduct experiments in conducive environments to gather accurate data. DNA methylation was the first known epigenetic modification. It is well-known how DNA methylation affects biology and the processes that control these changes. The functional roles of DNA methylation in cell differentiation, X chromosome inactivation, gene regulation, and epigenetic control were first proposed in 1975. Researchers are currently investigating the connections between other epigenetic markers, such as changes to the histone structure, and DNA methylation. A lot of labs look into and document variations in DNA methylation patterns as a contributing factor to or outcome of illnesses like cancer. Numerous methods have been developed to look into the genome-wide DNA methylation patterns.

Genome-wide analytic tools have been developed, though, by combining them with extensive cloning and sequencing efforts or combining them with microarray platforms. Most of the techniques were initially applied to single gene methylation changes detection. Often used on microarray platforms, promoter or CpG island arrays allow for the examination of important regulatory regions. One can study specific chromosome regions or the entire genome with arrays. Most methylation analysis techniques fall into numerous distinct classes based on their underlying principles. Multiple techniques are based on restriction endonucleases with altered sensitivity to methylation cytosine residues in the cleavage site. The restriction endonuclease digestion pattern is influenced by the cleavage site methylation state, and this pattern in turn reflects the methylation profiles of the corresponding chromosomal region. Proteins or antibodies that bind to methylated DNA are the basis of alternative techniques. Bisulfite sequencing methods are now being developed at high throughput; these methods are often required for resolution at the single nucleotide level.

Metabolism-sensitive restriction fingerprinting Metabolism-sensitive restriction fingerprinting (MSRF) was one of the first methods used by researchers to identify numerous de novo methylated or unmethylated sequences present in diverse tissue samples. For example, tissue-derived DNA samples are initially digested with MseI, a restriction enzyme that cleaves almost never within CpG islands. Sequencing and cloning of the amplified bands yields the information needed for target gene identification and database searches. Because of the high demand for gel electrophoresis, MSRF usage is currently limited. DNA regions with varying levels of methylation between two samples can be identified using a genome-wide analytical method known as Methylation-Sensitive Representational Difference Analysis (MS-RDA). Among its main characteristics is the enrichment of the unmethylated CpG-rich region of the genome by MS-RDA. DNA fragments from genomic DNA are amplified by MS-RDA using a methylation-sensitive restriction enzyme and a universal adaptor primer. One amplicon contains differentially methylated DNA fragments from the samples, while the other will not. Genomic subtraction is performed using the two amplicons, and the resulting fragments are cloned and sequenced. Restriction enzyme-based techniques also include differential methylation hybridization (DMH), which allows the simultaneous assessment of the methylation state of a large number of CpG-island sites. After ligation of the linker, methylation-sensitive enzymes are used to digest the genomic DNA from the target tissues, and the resulting products are used as templates for PCR on high-density arrays.

Methylated CpG Island Amplification Coupled Microarray (MCAM) is based on methylation-sensitive restriction enzymes and PCR with flanking primers, followed by sequence analysis of microarray probing. A powerful technique for concurrently detecting regions of the genome with differential methylation is MCAM. A methylation-sensitive enzyme is first used to cut the genomic DNA, resulting in blunt ends that aid in the removal of unmethylated sequences from the adaptor ligation process. The other enzyme, XmaI, creates sticky ends and recognizes the same sequence regardless of methylation status, enabling a correct adaptor ligation. After the methylated fraction has been amplified by PCR, fluorescently tagged, and hybridized, it is placed onto a microarray platform. Both normal and disease-derived tissue can readily be examined for a range of disease-related DNA methylation changes using MCAM. An alternative to the current methylation microarray techniques is the methylation-dependent restriction enzyme McrBC, which only cleaves methylated DNA. The precise cleavage point of this enzyme does not coincide with the methylation site, but the method is still effective for genome-scale DNA methylation analysis. As the foundation of sodium bisulfite sequencing, uracil is produced when sodium bisulfite deaminates cytosine. However, 5-methylcytosine is resistant to this process. Bisulfite sequencing allows for single-base resolution DNA methylation pattern analysis. The profiling of DNA methylation at specific loci or along significant chromosomal segments is based on the sequencing of PCR products of bisulfite-treated DNA. This method should ideally be used for full genome methylation profiling. Except for the few remaining methylated cytosines, sodium bisulfite treatment of DNA essentially divides the genome into three bases (A, G, and U/T). Many medium to high-throughput techniques have been developed to evaluate the methylation profiles of multiple genes at once after the conversion of genomic DNA with sodium bisulfite. In the mass spectrometry-based MALDI-TOF MS assay, the bisulfite-pretreated genomic DNA is utilized as a template for the PCR amplification of particular genes. One of the PCR primers is purposefully linked to the 5′ end by the T7 RNA polymerase promoter. The amplicon is transformed into single-stranded RNA *in vitro* using this promoter, and ribonuclease A is then used to digest the RNA at the uracil residues. Several genes can have direct quantitative methylation data simultaneously provided by a medium-throughput technique called MALDI-TOF MS. Typically, the number of multi-well formats used for the analysis is limited by the number of gene targets amplified and assessed in each run. Luminometric detection of the pyrophosphate produced during nucleotide incorporation is used in the pyrosequencing method of sequencing-by-synthesis. Bisulfite-converted DNA has been used as a template to show the feasibility and reliability of this method for the quantification of methylation at individual CpGs within PCR fragments. PCR is used with primers that match the DNA sequence that underwent bisulfite conversion to amplify a target region up to 350 kb. One of the two amplification primers has a biotin residue at the 5′ end to help with the purification of single-stranded templates. The template is hybridized with an internal pyrosequencing primer that is complementary to the single-stranded template. This quantitative approach is limited by the need to insert the primers into regions lacking CpG sites. The use of an antibody specific to 5-methylcytosine has facilitated the immunoprecipitation of highly methylated DNA sequences. After the fragments formed by the sonication of the genomic DNA are treated with an anti-5mC antibody, the methylated fraction is extracted from the reaction using Protein A/G beads. Large regions of repetitive DNA elements, such as transposable elements and satellite sequence clusters at telomeres, pericentric loci, and centromeres, comprise the majority of constitutive heterochromatin. Transposable elements are widely found in the human genome and are very mutagenic because they can be inserted into genes that code for proteins, breaking chromosomes and encouraging genome rearrangement. The constitutive heterochromatin that H3K9me3 maintains is crucial for preserving genomic integrity because it inhibits incorrect chromosome segregation, recombination, and DNA replication. Instead, facultative heterochromatin is primarily found at regulated loci where the chromatin state can change in response to cellular signals. The primary locations of facultative heterochromatin are the rRNA gene and the transcriptionally active region of euchromatin.

Did you get it?

1. **What do you understand by drugs?**
2. **Learn about enzymes**

FOCUS ON CAREER

- **Bioinformatician:** As a bioinformatician, individuals apply computational and statistical methods to analyze and interpret biological data, including genomic, transcriptomic, proteomic, and metabolomic data. Here's a detailed overview of careers in bioinformatics:
- **Bioinformatics Scientist:** They conduct research to develop algorithms, software tools, and databases for analyzing biological data. They collaborate with biologists, geneticists, and clinicians to design computational models, implement bioinformatics pipelines, and interpret experimental results. Bioinformatics scientists work in academic research labs, pharmaceutical companies, biotechnology firms, government agencies, and bioinformatics service providers.
- **Computational Biologist:** They use computational techniques to study biological systems, analyze large-scale data sets, and model biological processes. They apply mathematical modeling, machine learning, and network analysis methods to investigate genetic regulation, protein interactions, metabolic pathways, and evolutionary dynamics. Computational biologists work in research institutions, universities, and biotech companies, contributing to basic research and applied science.
- **Genomic Data Analyst:** These specialize in analyzing DNA sequencing data to identify genetic variants, characterize genomic features, and interpret biological insights. They use bioinformatics tools and genomic databases to analyze next-generation sequencing (NGS) data, perform variant calling, annotate genomic variants, and assess their functional impact. Genomic data analysts work in genome centers, clinical laboratories, and research consortia, supporting genomic research and precision medicine initiatives.
- **Clinical Bioinformatician:** These apply bioinformatics methods to analyze patient-derived genomic and clinical data for diagnosis, prognosis, and treatment of genetic diseases and cancer. They develop bioinformatics pipelines, interpret genomic test results, and provide insights into personalized treatment options based on molecular profiling. Clinical bioinformaticians work in hospitals, diagnostic laboratories, and precision medicine programs, collaborating with healthcare professionals to integrate genomic data into clinical practice.
- **Pharmaceutical/Biotech Bioinformatician:** These support drug discovery and development efforts by analyzing omics data to identify drug targets, predict drug responses, and optimize therapeutic interventions. They apply bioinformatics approaches to analyze high-throughput screening data, structure-activity relationships, and pharmacogenomic data to accelerate drug discovery pipelines. Pharmaceutical/biotech bioinformaticians work in drug discovery departments, research institutes, and contract research organizations (CROs).
- **Bioinformatics Consultant:** These provide expertise and support to research groups, companies, and organizations in designing and implementing bioinformatics solutions. They offer services such as data analysis, software

development, training, and project management to address specific research questions or computational challenges. Bioinformatics consultants may work independently as freelancers or as part of consulting firms specializing in bioinformatics services.

- **Academic Educator/Researcher:** Academic bioinformaticians teach bioinformatics courses, mentor graduate students, and conduct research in universities and academic institutions. They develop curriculum materials, design bioinformatics training programs, and collaborate on interdisciplinary research projects spanning biology, computer science, and statistics. Academic bioinformaticians contribute to advancing bioinformatics education and training, fostering the next generation of computational biologists and bioinformatics professionals.
- These are just a few examples of career paths available to bioinformaticians. With the rapid growth of biological data and the increasing demand for computational expertise in the life sciences, bioinformatics offers diverse opportunities for individuals interested in applying computational methods to address biological questions and solve real-world challenges.

CASE STUDY

EXPLORING GENE EXPRESSION: A CASE STUDY ON CYSTIC FIBROSIS

Introduction: Cystic fibrosis (CF) is a genetic disorder caused by mutations in the CFTR gene, leading to abnormal function or production of the cystic fibrosis transmembrane conductance regulator protein. This case study investigates the molecular mechanisms underlying gene expression in CF, its impact on cellular function, and therapeutic approaches for managing the condition.

Case Presentation: Emily, a 10-year-old girl, presents with recurrent respiratory infections, chronic cough, and poor weight gain. Further evaluation reveals salty-tasting skin, nasal polyps, and a family history of CF in her older brother. Genetic testing confirms a homozygous mutation in the CFTR gene, leading to a diagnosis of cystic fibrosis. Emily's parents are devastated by the news and seek guidance on managing her condition.

Discussion:

- **Gene Expression and CFTR Protein:** The CFTR gene encodes the CFTR protein, which functions as a chloride ion channel in epithelial cells lining various organs, including the lungs, pancreas, and sweat glands. Mutations in the CFTR gene disrupt the production or function of the CFTR protein, impairing chloride ion transport across cell membranes. This leads to thick, sticky mucus buildup in the respiratory, digestive, and reproductive systems, causing recurrent infections, digestive problems, and infertility in affected individuals.

- **Molecular Mechanisms:** Gene expression of the CFTR gene involves multiple steps, including transcription, mRNA processing, translation, and protein folding. Mutations in the CFTR gene can occur at various stages of gene expression, affecting mRNA stability, protein synthesis, or protein function. Common mutations in CFTR, such as ΔF508, result in defective protein folding and trafficking, leading to reduced CFTR channel activity on the cell surface.
- **Cellular Consequences:** Dysfunction of the CFTR protein disrupts ion and fluid balance in epithelial tissues, leading to several cellular consequences in CF:
  - Impaired mucociliary clearance: Thickened mucus accumulates in the airways, impairing lung function and predisposing to respiratory infections.
  - Pancreatic insufficiency: Reduced CFTR function in pancreatic duct cells leads to decreased bicarbonate secretion and impaired enzyme delivery, resulting in malabsorption and nutritional deficiencies.
  - Sweat electrolyte imbalance: Defective chloride ion transport in sweat glands causes increased salt concentration in sweat, a hallmark feature of CF.
- **Therapeutic Approaches:** The management of CF involves a multidisciplinary approach aimed at addressing respiratory, nutritional, and psychosocial needs. Therapeutic interventions include:
  - Airway clearance techniques and inhaled medications to improve lung function and prevent infections.
  - Pancreatic enzyme replacement therapy and nutritional supplementation to optimize digestion and growth.
  - CFTR modulator therapies target specific mutations to enhance CFTR function or correct protein folding defects.
  - Supportive care and counseling to address psychosocial challenges and improve the quality of life for patients and families.

Conclusion: Cystic fibrosis exemplifies the intricate relationship between gene expression, protein function, and cellular physiology. Understanding the molecular mechanisms underlying CF provides insights into its pathogenesis and informs the development of targeted therapies aimed at correcting CFTR dysfunction. Advances in genetic research and personalized medicine continue to drive progress in CF management, offering hope for improved outcomes and quality of life for individuals living with this chronic genetic disorder.

SUMMARY

- Gene expression is the process by which a gene's information is used to create a functional product, usually a functional RNA molecule or protein.
- Gene regulation is essential for viruses, prokaryotes and eukaryotes as it increases the versatility and adaptability of an organism by allowing the cell to express protein when needed.
- Gene expression regulation preserves both space and energy. Requiring the genes to be activated only when necessary results in greater energy efficiency.
- Translational regulation takes place during the synthesis of proteins, when the mRNA's nucleotide sequence is decoded to put together a particular arrangement of amino acids to form a polypeptide chain.
- Post-translational regulation pertains to the alteration and handling of proteins subsequent to their synthesis. This covers a broad variety of covalent modifications, including glycosylation, ubiquitination, acetylation, phosphorylation, and proteolytic cleavage.
- The stability of the mRNA in the cytoplasm is a form of post-transcriptional control. An mRNA must take longer to translate and produce more protein the longer it is present in the cytoplasm.
- Post-translational regulation is the last stage of gene expression regulation in eukaryotes. This kind of control entails altering the protein after it is produced in order to change its activity.
- Chromatin modifications are another crucial epigenetic mechanism that impacts gene expression along with numerous other biological functions.
- Gene or region-specific DNA methylation analysis is necessary in a number of circumstances, and numerous methods are available.

MULTIPLE CHOICE QUESTIONS

1. **What is the first step in gene expression?**
   a. Translation
   b. Transcription
   c. mRNA processing
   d. Protein folding
2. **Which of the following molecules carries the genetic information from DNA to the ribosome for protein synthesis?**
   a. tRNA
   b. mRNA
   c. rRNA
   d. DNA polymerase

3. **In eukaryotic cells, where does mRNA processing occur?**
   a. Nucleus
   b. Cytoplasm
   c. Golgi apparatus
   d. Endoplasmic reticulum
4. **Which of the following enzymes catalyzes the synthesis of mRNA during transcription?**
   a. DNA ligase
   b. RNA polymerase
   c. DNA polymerase
   d. Helicase
5. **Which of the following is a regulatory protein that binds to DNA and controls the initiation of transcription?**
   a. Ribosome
   b. Promoter
   c. Transcription factor
   d. tRNA

REVIEW QUESTIONS

1. What is gene expression, and why is it essential for living organisms?
2. Describe the central dogma of molecular biology and the processes involved in gene expression.
3. What are the key differences between transcription and translation?
4. Explain how RNA polymerase recognizes and binds to specific regions of DNA during transcription.
5. What are the roles of transcription factors in gene regulation?

Answers to Multiple Questions

1. (b); 2. (b); 3. (a); 4. (b); 5. (c)

REFERENCES

1. Cheadle, C., Fan, J., Cho-Chung, Y. S., Werner, T., Ray, J., Do, L., et al. (2005). Control of gene expression during T cell activation: Alternate regulation of mRNA transcription and mRNA stability. *BMC Genomics*, *6*, 75.
2. Jabbari, K., & Bernardi, G. (2004). Cytosine methylation and CpG, TpG (CpA) and TpA frequencies. *Gene*, *333*, 143–149.

3. Latchman, D. S. (2005). *Gene Regulation: A Eukaryotic Perspective*. Psychology Press.
4. Levine, A., Huang, Y., Drisaldi, B., Griffin, E. A., Pollak, D. D., Xu, S., et al. (2011). Molecular mechanism for a gateway drug: Epigenetic changes initiated by nicotine prime gene expression by cocaine. *Science Translational Medicine*, *3*(107), 107ra109.
5. Remenyi, A., Scholer, H., & Wilmanns, M. (2004). Combinatorial control of gene expression. *Nature Structural and Molecular Biology*, *11*, 812–815.
6. Schwanekamp, J. A., Sartor, M. A., Karyala, S., Halbleib, D., Medvedovic, M., & Tomlinson, C. R. (2006). Genome-wide analyses show that nuclear and cytoplasmic RNA levels are differentially affected by dioxin. *Biochimica et Biophysica Acta (BBA)–Gene Structure and Expression*, *1759*(8, 9), 388–402.
7. Sproul, D., Gilbert, N., & Bickmore, W. (2005). The role of chromatin structure in regulating the expression of clustered genes. *Nature Reviews Genetics*, *6*, 775–781.
8. Statello, L., Guo, C. J., Chen, L. L., & Huarte, M. (2021). Gene regulation by long non-coding RNAs and its biological functions. *Nature Reviews Molecular Cell Biology*, *22*(2), 96–118.
9. Yang, X., Han, H., De Carvalho, D. D., Lay, F. D., Jones, P. A., & Liang, G. (2014). Gene body methylation can alter gene expression and is a therapeutic target in cancer. *Cancer Cell*, *26*(4), 577–590.

CHAPTER

5

Population Genetics

LEARNING OBJECTIVES

After studying this chapter, you will be able to:

- Discuss the concept of population genetics;
- Learn about Hardy-Weinberg law;
- Explain the population-genetic models of evolution;
- Focus on genetic-automatic processes.

INTRODUCTORY EXAMPLE

Fundamentals of Population Genetics:

Population genetics is the field of biology that studies the genetic composition and evolutionary dynamics of populations. By examining patterns of genetic variation within and between populations, population geneticists gain insights into the forces that shape the genetic diversity of species over time. To illustrate the fundamental concepts of population genetics, let's consider a hypothetical example involving a population of fictional creatures called "Trilobites." Imagine a population of Trilobites inhabiting a vast, diverse ecosystem. Trilobites are small, beetle-like creatures with a variety of physical traits, including body size, coloration, and appendage length. Within this population, genetic variation exists for each of these traits due to differences in the DNA sequences that encode them.

One of the key concepts in population genetics is allele frequency, which refers to the proportion of a particular allele—the different forms of a gene—within a population's gene pool. For instance, let's focus on a gene that influences body size in Trilobites. This gene has two common alleles: "big" and "small." In our hypothetical population, 60% of Trilobites carry the "big" allele, while 40% carry the "small" allele. These allele frequencies provide insights into the genetic makeup of the population and how traits may be inherited from one generation to the next. Another important concept is genetic drift, which describes the random fluctuations in allele frequencies that occur over time due to chance events. In our Trilobite population, imagine a severe storm that sweeps through the ecosystem, leading to the accidental deaths of many individuals. Due to this catastrophe, the allele frequencies in the surviving population may shift. For example, if more Trilobites with the "big" allele happen to perish in the storm, the frequency of the "big" allele may decrease in the remaining population. Genetic drift is particularly influential in small populations where chance events can have a significant impact on allele frequencies. In our example, if a small group of Trilobites becomes isolated from the main population—for instance, on a remote island—genetic drift may cause allele frequencies in the isolated population to diverge from those in the larger population over time. Natural selection is another driving force in population genetics. It acts on heritable traits, favoring those that confer a reproductive advantage in a given environment. For instance, suppose there is a change in the environment of Trilobites, such as an increase in predation by larger organisms. Trilobites with traits that enhance their ability to evade predators—perhaps those with darker coloration camouflaging them better—may have higher survival rates and, consequently, higher reproductive success. Over generations, natural selection can lead to changes in allele frequencies as advantageous traits become more common in the population.

Migration, or gene flow, is also essential in population genetics. It refers to the movement of individuals or alleles between populations, which can introduce new genetic variation or alter existing allele frequencies. For instance, if Trilobites from different regions of the ecosystem migrate and interbreed, they may exchange genetic material, leading to a mixing of traits and allele frequencies between populations.

UNIT INTRODUCTION

The area of genetics known as population genetics focuses on how genetic variation is distributed within populations and how it changes through time. In addition to examining the variables influencing these processes, it looks at how genetic variation develops, persists, and changes within populations. Elemental frequencies, or the percentage of distinct alleles (different versions of a gene) in a population, are fundamental to the study of population genetics. A key idea in population genetics, the Hardy-Weinberg equilibrium, describes how allele frequencies in idealized populations can remain stable under specific circumstances and serve as a baseline against which genetic variation in the real world can be measured. Gene flow, natural selection, genetic drift, mutation, and recombination are important evolutionary processes that mold genetic variation patterns within populations. Natural selection modifies allele frequencies over generations by favoring individuals with advantageous alleles. This process is applied to heritable traits. Small populations are especially affected by genetic drift, which is the random variation in allele frequencies brought on by chance events. Gene flow refers to the transfer of genes from one population to another, resulting in the exchange of genetic material and the standardization of allele frequencies. Mutation creates new genetic variation in populations, whereas, during meiosis, recombination creates unique allele combinations on chromosomes.

Understanding population genetics can help one better understand a range of biological phenomena, such as genetic diversity, speciation, adaptation, and disease susceptibility. It is useful in a variety of industries, including forensics, agriculture, medicine, and conservation biology. The development of personalized medicine approaches, for instance, can benefit from an understanding of the genetic basis of disease susceptibility, and conservation efforts can be guided by population genetic analyzes that identify genetically distinct populations and prioritize conservation strategies.

Overall, population genetics provides a strong framework for comprehending how populations respond to environmental stresses and how their genetic dynamics change. This framework sheds light on the mechanisms underlying evolution and the diversity of life on Earth.

5.1. CONCEPT OF POPULATION GENETICS

Learning Objectives

- What is population?
- Discuss the properties of populations;
- Explain the population and environment.

As a branch of evolutionary biology, population genetics is the study of genetic variations both within and between populations. Research in this area of biology looks at population structure, speciation, and adaptation.

Population genetics was a vital ingredient in the emergence of the modern evolutionary synthesis. Its primary founders were Sewall Wright, J. B. S. Haldane, and Ronald Fisher, who also laid the foundations for the related discipline of quantitative genetics. Traditionally a highly mathematical discipline, modern population genetics encompasses theoretical, lab, and field work. Population genetic models are used both for statistical inference from DNA sequence data and for proof/disproof of concept.

Population genetics today is distinguished from more phenotypic, contemporary approaches to modeling evolution, such as adaptive dynamics and evolutionary game theory, by its focus on genetic phenomena such as dominance, epistasis, and the extent to which genetic recombination disrupts linkage disequilibrium. It can therefore be used in conjunction with population genomics data.

Analyzing the interactions between predictable, "deterministic" evolutionary forces and random, unpredictable, "stochastic" forces is the essence of population genetics. Allele frequencies are pushed in one direction (up, down, or towards the middle) by the deterministic forces, which are commonly referred to as "linear pressures." Meiotic drive, which is the unequal transmission of some alleles [a form of selection], selection, mutation, gene flow, and non-random mating are significant examples of these forces. Genetic drift, which results from the random sampling of individuals (and genes) in small populations, is the main stochastic evolutionary force. Furthermore, to establish the frequencies of alleles and genotypes in populations, deterministic forces may operate in tandem with or against genetic drift (e.g., drift tends to keep populations distinct, whereas gene flow tends to homogenize them. Because of this, the interaction of these forces is really what we are interested in (more on this in a later lecture), but for the time being, we will focus on analyzing each force separately due to the mathematical complexity of this concept. Many consider population genetics to be the theoretical foundation of contemporary Darwinism since it is closely related to the study of evolution and natural selection. This is since one of the most significant variables influencing a population's genetic makeup is natural selection. When certain variations in a population outproduce others because they are "fitter," or more adapted to the environment, natural selection takes place. Assuming that genetic differences account for a portion of the fitness disparities, this will gradually change the genetic composition of the population. Therefore, population geneticists aim to clarify the evolutionary process and enable the quantitatively precise investigation of the implications of various evolutionary hypotheses by examining formal models of gene frequency change.

Remember

Since populations consist of individuals, who occur in communities together with other organisms, populations should be examined at a different organizational level than both individual organisms and communities.

5.1.1. What Is Population?

Like any other ecological entity, the term "population" needs to be defined. This serves two functions: (a) distinguishing populations from other entities that are not populations; (b) distinguishing between various populations. The most commonly used definition of a population is a group of individuals belonging to the same species that coexist in a specific location and time. This is an issue of observational "focus," whereby only the aggregate numbers of the group of people and their interactions with one another are taken into account. From a population perspective, individual experiences are considered "detail," while interactions with other organisms are considered to be a part of the "background." Although they both describe phenomena that can be studied with a different focus, they are both crucial for understanding populations.

The definition's secondary goal is more dubious. This is because all of the definition's elements—individual, species, space, and time—are extremely ambiguous. This is an issue that is specific to biological systems and illustrates how varied and dynamic they are. Expert knowledge of the target organisms and the application of ad hoc definitions tailored to the particular query and the distinctive characteristics of the organisms provide the answer.

Problems often arise when defining where in space a population is. For some organisms that are fixed to a certain place or move only short distances, we can know exactly where the individuals are, but not where the boundaries of the population are. In these cases, decisions are often taken arbitrarily based on sampling considerations, or based on landscape boundaries (a recognizable patch, a woodlot, etc.).

Though it is the most concrete entity in ecology, the term "individual" is the most ambiguous in the definition. Because individuals in unitary organisms are distinct both genetically and physiologically, identifying them is typically not a problem. Vegetative propagation is a method used by many plant and invertebrate species in addition to sexual reproduction for creating new individuals. A single genetically distinct individual (the genet) gives rise to new functional modules (ramets) in these modular organisms. Not genetically distinct, but countable as individuals if the ramets are physiologically independent. The only difference between this and a plant's internodes and branches is the absence of the connections.

The term "species" may also cause issues if different taxonomy definitions of the term are used, which could split one species or combine two, or if the biological species concept—which holds that all individuals must be able to mate—is not relevant. This is the case in situations where there is either cleistogamy—sexual reproduction by self-pollination without interindividual crossing in many plants, such as wild barley—or fragmentation—a type of vegetative propagation common in flatworms, for example.

Did you know?

Population genetics was a vital part in the emergence of the modern evolutionary synthesis. Its primary founders were Sewall Wright, J. B. S. Haldane, and Ronald Fisher, who also laid the foundations for the related discipline of quantitative genetics.

5.1.2. Properties of Populations

The question "what determines the structure and functioning of a population?" suggests that our goal is to learn about the variables that affect a population and how they affect it. To address these questions, we need to identify the characteristics of populations that are characterized and the types of factors that can affect them. First and foremost, there are significant differences between populations of different species; even within the same species, there can be significant differences between two populations. Even so, each population shares certain characteristics, albeit in varying amounts and forms.

These properties are:

1. **Dimensions:** population density, abundance or size, based on the number of individuals (per unit area, or in the entire population);
2. **Composition:** the proportions of individuals with different states (gender, age, developmental stage, or size);
3. **Dynamics:** changes in time of the dimensions and composition of a population; this includes the rate of change of the dimensions (population growth rate) and changes in the distributions of individuals' states.

Population characteristics are constantly changing, with a wide range of change rates. Individuals change and are replaced even in populations with constant densities. They are born, develop, procreate—or not—and then pass away. These individual state changes can be summed up as rates of change, like the birth and death rates, or as the percentage of people who reach a specific size over time, among other metrics. The dynamics, composition, and dimensions of the population are directly impacted by these demographic processes, which also contribute to its variation or absence.

Demographic processes are abstractions that come from real shifts in people's conditions. They are computed as the percentage of people who make it to the next stage from a given stage. Based on their genotype, history, and overall health, each person experiences change at a different pace. Vital rates are averages of demographic processes (based on age, size, or stage) over groups of people. Vital rates encompass growth, survival, fecundity, and birth rate. The population's life cycle is determined by the conditions of each individual and the rates at which those states connect.

5.1.3. Population and Environment

Vital rates, which indicate how individuals react to environmental variation, are the first point of difference between two populations of the same species. When comparing the food availability of a population of a specific animal species in rich and poor habitats, for instance, we typically find that the richer habitat has a higher population growth rate than the poorer habitat. This is because populations that have access to more food tend to survive better, grow more quickly, reach reproductive maturity more quickly, and produce more offspring overall.

Environmental factors that affect demographic processes are very diverse and can be divided in many ways. Categories are not exclusive: some factors may appear in more than one category. The following division is based on the properties of the factors and their effects on the organisms.

1. **Environmental Conditions:** States of the environment that affect organisms, but are not consumed by them: pH, temperature, air humidity (including their variation).
2. **Resources:** Physical commodities (material or energy in any form, including other organisms), which are supplied by the environment and consumed by the organism; resources are at least as diverse as the organisms that require them.
3. **Structures:** Availability of sites where individuals can perform their essential functions (shelter, nesting, foraging, mating, and recruitment sites); sites can be viewed as a structural sort of resource.
4. **Interactions with Other Organisms:** All relationships with other organisms except consuming another as a resource; interactions include (the passive side of) predation, herbivory and parasitism, and competition for resources and structures; besides negative interactions, they can be positive as well (facilitation, mutualism).
5. **Internal Regulation:** Effects of the population itself (or some of its members) on the way environmental factors affect the organisms. (Strictly speaking, this is not an environmental factor, although for each member of a population, the other members are part of its environment.) Processes affected this way are called density-dependent. Density-dependence can be positive or negative, depending on whether the density of the population reduces or enlarges the effect of the environmental factor. Intraspecific competition is a negatively density-dependent process and operates on resource acquisition.

5.1.4. Explaining Population Phenomena

Figure 5.1 illustrates the logical relationships between environmental factors and population characteristics and demography. One or more demographic processes within the population are influenced by a factor. This suggests that a vital rate (survival, growth, and reproduction) responds to a change in the factor's level by increasing or decreasing. This is another way of saying that the factor affects people and modifies their capacity for growth, survival, and procreation. Lastly, the dynamics and composition of the population as a whole are shaped by the demographic processes.

Keyword

Environmental Factor – It is any factor, abiotic or biotic, that influences living organisms. Abiotic factors include ambient temperature, amount of sunlight, and pH of the water soil in which an organism lives.

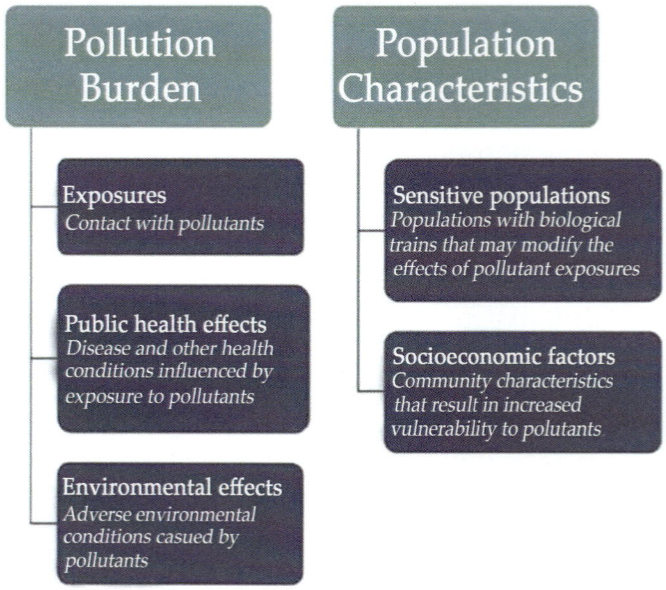

Figure 5.1. *Relations between environmental factors, the demography and properties of a population, and internal regulation.*

Source: Meehan, A. L., Faust, J. B., Cushing, L., Zeise, L., &Alexeeff, G. V., (2012). Methodological considerations in screening for cumulative environmental health impacts: Lessons learned from a pilot study in California. International Journal of Environmental Research and Public Health, 9(9), 3069–3084. https://doi.org/10.3390/ijerph9093069.

5.1.5. Population Research Methods

The most precise method of estimating a population's size is to count every member of the population. This strategy is typically impractical, though, particularly in cases of high populations or vast habitats. Rather, by counting individuals within a specific volume or area that is a part of the population's habitat, scientists can learn more about populations through sampling. Scientists can determine population size and density for the entire population by analyzing sample data. To sample populations, one can employ a number of techniques. The quadrat method is typically used by scientists to estimate the populations of sessile or slowly moving organisms. A square that contains a section of a habitat is called a quadrat. The space can be marked off by putting a metal, plastic, or wood square on the ground or by staking it out with sticks and string.

Multiple quadrat samples are typically taken at random intervals or along a transect in a representative habitat for a field study. The number of people who fall inside the quadrat boundaries is counted by the researchers after the quadrats are placed. The number of samples, the size of the quadrats, the type of organism, its spatial distribution, and other details are determined by the researcher. For sampling daffodils, a 1 m^2 quadrat could be appropriate. Giant redwoods are larger and live further apart from each other, so a larger quadrat, such as 100 m^2, would be necessary. The correct quadrat size ensures counts of enough individuals to get a sample representative of the entire habitat.

The number of individuals in the population can be estimated as the total population size by scientists using the ratio of marked to unmarked individuals. This method works on the assumption that the more individuals in the population, who have mixed with more untagged ones, the smaller the percentage of tagged organisms that will be recaptured. For example, if 80 deer are captured, tagged, and released into the forest, and later 100 deer are captured with 20 of them are already marked, we can determine the population size (N) using the following equation:

$$\frac{number\ marked\ in\ first\ catch \times total\ number\ of\ \sec ond\ catch}{number\ of\ marked\ recaptures\ in \sec ond\ catch}$$

Plugging the example data into the equation, the calculation gives an estimated total population size of 400.

$$\frac{80 \times 100}{20} = 400$$

Using the example data, if only 10 already-marked deer had been recaptured, the calculated total population size would be 800.

Did you get it?

1. What do you understand by population phenomena.
2. Describe the population research methods.

5.2. HARDY-WEINBERG LAW

Learning Objectives

- Discuss about Hardy-Weinberg equilibrium;
- Focus on genetic drift, gene flow, and natural selection.

In any intrabreeding population, each pair of alleles (A, a) has quite regular correlations of genotypes (AA, Aa, and aa) described by Hardy and Weinberg (1908) and is known as the Hardy-Weinberg law. This law is formulated as follows. If in a large panmictic population, there are alternative forms of an autosome gene, without mutation, selection, and differential migration, the initial numerical correlations of these alleles (P_1, P_2, P_3... P_n) will be the same from generation to generation; and after one change of generations, the frequency of genotypes will again obtain an equilibrium state. Equilibrium frequencies of genotypes correspond to the terms of the expression obtained by squaring the formula (P_1+P_2+P_3+... P_n).

Genetic inertia is a characteristic of Mendelian populations under these circumstances. The population's genetic makeup does not alter if mutation, selection, specific parent pair modifications, and population size reductions do not result in imbalance.

Though these conditions are unachievable in natural populations, an analysis of the allele distribution in an infinitely large population under these conditions reveals the main trends in genotype distribution in natural populations and offers a foundation for comprehending the role of disturbances brought on by the factors listed above.

The Hardy-Weinberg law describes the regular ratio of homozygotes and heterozygotes for any random mating population. This regular ratio forms the basis of the concentration of genes in the population. Gene concentration is related to the frequency of genes in the population. If the amount of the dominant gene A in a population is P, its recessive allele in the same population has naturally a concentration equal to 1 – p or q. Equal distribution of genes among males and females results in the fact that among ova as well as among spermatozoa, we find gametes out of which p bear A genes and q gametes bear a genes (Table 5.1).

Table 5.1. The Most Probable Recombination of Male and Female Gametes When Intrabreeding

Sperm	Ova	
	pA	Qa
pA	p^2AA	pqAa
qa	pqAa	q^2aa

Source: https://www.fao.org/3/b3310e/B3310E07.htm.

We obtain $p^2AA + 2qpAa + q^2aa$. This expression is the formula of the Hardy-Weinberg law. According to this formula, the amount of heterozygotes is equal to the double product of the concentrations of both alleles.

Genetic concentration determination is challenging when working with real-world populations because many genes turn out to be recessive and have very low proportions in populations.

For instance, albino frequency, i.e., frequency of white-haired and red-eyed specimens, averages 1: 20,000.

According to genetic theory, human albinos are homozygotes of the recessive gene d, which results in a total lack of pigment. There are heterozygous individuals with the normal allele Dd and homozygous individuals with the normal allele DD among normal individuals. We are able to ascertain the gene D concentration and the number of heterozygous individuals whose progeny will be albinos. The concentration of the gene d (albinism) is:

$$\mathrm{d} = \sqrt{\frac{1}{20000}} = \frac{1}{141};$$

The concentration of the gene D (with pigment) is equal to:

$$D = p = 1 - \frac{1}{141} = \frac{140}{141};$$

The frequency of homozygotes of the dominant allele (DD) is equal to

$$DD = P^2 \left(\frac{140}{141}\right)^2 =\sim \frac{68}{70}$$

The frequency of phenotypically equal heterozygotes (Dd) is equal to

$$Dd = 2pq = 2x\frac{140}{141}x\frac{1}{141} =\sim \frac{1}{70}$$

Therefore, the frequency of heterozygotes (Dd) of people who have albino offspring averages 1:70, despite the fact that albinos (homozygotes dd) occur in human populations at a frequency of 1:20 000. This demonstrates that a population needs a comparatively large number of heterozygotes for their emergence, even in the case of a small number of homozygotes. The likelihood of coming into contact with heterozygotes in family marriages, which can lead to the emergence of distinct defective genes, increases sharply.

Keyword

Genetic Structure – It refers to any pattern in the genetic makeup of individuals within a population.

The concentrations of the recessive allele (q) and the dominant allele (p) determine the genotype distribution in a population. Variations in the allele concentrations around 0.5 do not significantly impact the number of heterozygotes. Heterozygotes' frequencies don't start to change significantly until there is a significant variance in the allele concentrations.

It is highly likely that specimens with distinct genotypes will cross in a population. Variations in allele concentrations impact the likelihood of encountering specimens with distinct genetic structures. Obviously if in a population both males and females of genotype (AA) occur in the concentration p^2, the probability of their crossing is $p^2 \times p^2 = p^4$. In addition to the crossing of homozygotes with the dominant allele (AA × AA) and the recessive allele (aa × aa) four more types of crossing may occur: Aa × Aa; AA × Aa; aa × AA and Aa × aa. For each crossing-type, it is possible to calculate its probability for populations with different proportions of alleles. Illustrating the ratio of frequencies of different crossings in populations with concentrations varying from 0 to 1. We can see the pattern of these relations and notice that irrespective of the value of concentrations of genes, the frequency of meetings of heterozygotes (Aa × Aa) is always twice as high as the frequency of meetings of homozygotes (aa × aa, AA × AA).

When two independent alleles (A and a) are found in combinative variation along with a few other pairs of alleles (B and b), the law of probable distribution is demonstrated by three-term formulas for each independent allele. In this instance, even if a population of that kind were to lose its equilibrium state. Stabilizing crosses are those that change the population genotypes' previous equilibrium distribution into a new equilibrium state based on altered relative allele concentrations.

Naturally, none of the requirements required for the equilibrium law to be realized are met in real populations. It does not, however, imply that the significance of the Hardy-Weinberg law disappears. It is crucial to understand the genotype distribution formula in a population. By calculating the concentrations of various alleles, this formula can be analyzed to determine the genetic makeup of a population. Nevertheless, this equation is only valid for one frame—the current generation—of an ongoing movie that can be used to compare the dynamic evolution of successive generations.

This permanent development in the genetic structure of populations associated with isolations, migrations, mutations and selections is the internal side of all life processes of different units of the species in different parts of its habitat.

5.2.1. Hardy-Weinberg Equilibrium

According to the Hardy-Weinberg equilibrium, if there are no disruptive factors, genetic variation in a population will not change from generation to generation. Given a large population and random mating, the law predicts that allele frequencies and genotype frequencies will stay constant because they are in an equilibrium.

Mutations, natural selection, non-random mating, genetic drift, and gene flow are some of the factors that can upset the Hardy-Weinberg equilibrium. By introducing new alleles into a population, mutations, for example, upset the equilibrium of allele frequencies. In the same way, because they alter gene frequencies, natural selection, and nonrandom mating upset the Hardy-Weinberg equilibrium. This happens because the organisms that carry particular alleles benefit or suffer from their reproductive success. Genetic drift, which is the rise or fall in allele frequencies due to random variation and usually occurs in small populations, is another factor that can disturb this equilibrium. The Hardy-Weinberg equilibrium can also be changed by gene flow, which happens when individuals breed with each other to introduce new alleles into a population.

The Hardy-Weinberg equilibrium is rarely true in practice because all of these disruptive forces are frequently found in nature. Genetic variations in nature can therefore be quantified as deviations from the idealized state that the Hardy-Weinberg equilibrium describes.

The seven assumptions underlying Hardy–Weinberg equilibrium are as follows:

- Organisms are diploid;
- Only sexual reproduction occurs;
- Generations are nonoverlapping;
- Mating is random;
- Population size is infinitely large;
- Allele frequencies are equal in the sexes;
- There is no migration, gene flow, admixture, mutation or selection.

Violations of the Hardy–Weinberg assumptions can cause deviations from expectations. How this affects the population depends on the assumptions that are violated.

- **Random Mating:** The HWP states the population will have the given genotypic frequencies (called Hardy–Weinberg proportions) after a single generation of random mating within the population. When the random mating assumption is violated, the population will not have Hardy–Weinberg proportions. A common cause of non-random mating is inbreeding, which causes an increase in homozygosity for all genes.

  If a population violates one of the following four assumptions, the population may continue to have Hardy–Weinberg proportions each generation, but the allele frequencies will change over time.

- Selection, in general, causes allele frequencies to change, often quite rapidly. While directional selection eventually leads to the loss of all alleles except the favored one (unless one allele is dominant, in which case recessive alleles can survive at low frequencies), some forms of selection, such as balancing selection, lead to equilibrium without loss of alleles.
- Mutation will have a very subtle effect on allele frequencies through the introduction of new alleles into a population. Mutation rates are of the order 10^{-4} to 10^{-8}, and the change in allele frequency will be, at most, the same order. Recurrent mutation will maintain alleles in the population, even if there is strong selection against them.
- Migration genetically links two or more populations together. In general, allele frequencies will become more homogeneous among the populations. Some migration models inherently include nonrandom mating (Wahlund effect, for example). For those models, the Hardy–Weinberg proportions will normally not be valid.
- Small population size can cause a random change in allele frequencies. This is due to a sampling effect, and is called genetic drift. Sampling effects are most important when the allele is present in a small number of copies.

In real-world genotype data, deviations from Hardy–Weinberg Equilibrium may be a sign of genotyping error.

5.2.1.1. Sex Linkage

Where the A gene is sex-linked, the heterogametic sex (e.g., mammalian males; avian females) have only one copy of the gene (and are termed hemizygous), while the homogametic sex (e.g., human females) have two copies. The genotype frequencies at equilibrium are p and q for the heterogametic sex but p^2, $2pq$, and q^2 for the homogametic sex.

For example, in humans red–green colorblindness is an X-linked recessive trait. In western European males, the trait affects about 1 in 12, ($q = 0.083$) whereas it affects about 1 in 200 females (0.005, compared to $q^2 = 0.007$), very close to Hardy–Weinberg proportions.

As each son inherits its X chromosome from its mother, if a population of males and females with distinct allele frequencies in each subpopulation (males or females) is combined, the male population's allele frequency in the following generation will coincide with the female population's. Very quickly, the population converges on equilibrium.

5.2.2. Genetic Drift, Gene Flow, and Natural Selection

Natural selection, gene flow, and genetic drift are the three main processes that propel evolutionary change within populations, affecting the frequency and distribution of alleles over time:

1. **Genetic Drift:** The random fluctuation of allele frequencies in a population brought on by chance events is referred to as genetic drift. This phenomenon is especially

noticeable in small populations, where allele frequencies can be significantly impacted from generation to generation by random sampling effects. Over time, genetic drift tends to decrease genetic diversity and can result in allele loss or fixation, where one allele becomes the only allele present at a specific locus.

2. **Gene Flow:** This sometimes referred to as migration, is the transfer of genes from one population to another via individual migration. Depending on the amount and direction of migration, gene flow can either homogenize allele frequencies across populations or introduce new alleles into an existing population. Gene flow can sustain genetic diversity and counteract genetic drift in populations at high levels, while low levels can cause genetic cluster formation and population differentiation.
3. **Natural Selection:** It is the process through which certain heritable traits give rise to an advantage in reproduction, increasing their frequency in a population over the course of successive generations. Natural selection favors individuals possessing traits that increase their chances of surviving and procreating in a particular environment based on the variation that exists within a population. An organism's fitness—or its capacity to endure and procreate in its ecological niche—may be enhanced as a result of these adaptations. Different mechanisms, such as directional selection (favoring individuals at one extreme of a phenotypic trait), stabilizing selection (favoring individuals with intermediate phenotypes), and disruptive selection (favoring individuals at both extremes of a phenotypic trait), can allow natural selection to process information.

In order to shape the genetic makeup and evolutionary paths of populations, genetic drift, gene flow, and natural selection work together to great effect. In addition to addressing conservation and management challenges in the face of environmental change, understanding these mechanisms is crucial for clarifying patterns of genetic diversity, population structure, and adaptation in natural populations.

Did you get it?

1. **What do you mean by sex linkage?**

5.3. POPULATION-GENETIC MODELS OF EVOLUTION

Learning Objectives

- How to selection at one locus;
- Focus on selection-mutation balance;
- Learn about random drift;
- What do you mean by migration.

Any alteration in a population's genetic makeup over time is typically referred to as "evolution" by population geneticists. Natural selection, mutation, random genetic drift, and migration into or out of the population are the four processes that can cause such a shift.

5.3.1. Selection at One Locus

In a population, natural selection takes place when certain genotypic variants have an advantage over others in terms of reproduction or survival. One autosomal locus with two alleles, A_1 and A_2, is assumed in the most basic population-genetic model of natural selection. The three diploid genotypes, represented by the letters w_{11}, w_{12}, and w_{22}, respectively, A_1A_1, A_1A_2, and A_2A_2 have varying fitnesses. It is assumed that these fitnesses will remain unchanged over generations. This definition of genotype fitness can be summed up as the average number of viable gametes that an organism carrying that genotype passes on to the following generation. This number is determined by the organism's ability to survive, reproduce, and maintain fertility. Natural selection is going to happen if w_{11}, w_{12}, and w_{22} are not equal. This could result in a change in the genetic makeup of the population.

Assume that at first, i.e., the frequencies of the A_1 and A_2 alleles are p and q, respectively, before selection takes effect, with $p+ q = 1$. The zygote genotypes are in Hardy-Weinberg proportions before selection takes place. After reaching adulthood, the zygotes procreate and create a new generation of offspring zygotes. Calculating the frequencies of A_1 and A_2 in the second generation is our task; we will refer to them as p' and q', respectively, with $p' + q' = 1$.

According to the Hardy-Weinberg law, the genotypic frequencies at the zygotic stage for A_1A_1, A_1A_2, and A_2A_2 are p^2, 2pq, and q2, respectively. According to their respective fitnesses, the three genotypes generate successful gametes, i.e., according to w_{11}:w_{12}:w_{22}. The average fitness in the population is $\overline{w} = p^2w_{11} + 2pqw_{12} + q^2w_{22}$, so the total number of successful gametes produced is $N\overline{w}$, where N is the size of the population. An A_1A_1 organism will only produce A_1 gametes, an A_2A_2 organism will only produce A_2 gametes, and an A_1A_2 organism will produce A_1 and A_2 gametes in equal amounts, assuming that there is no mutation and that Mendel's law of segregation is valid. As a result, the frequency of the A_1 allele in the second generation during the zygotic stage and the proportion of A_1 gametes are as follows:

$$p' = [N p^2 w_{11} + ½ (N 2pq w_{12})] / N\overline{w} \\ = (p^2 w_{11} + pq w_{12}) / \overline{w} \quad (1)$$

Referred to as a "recurrence" equation, Eqn. (1) expresses the frequency of the A_1 allele in the second generation relative to its frequency in the first generation. This allows us to write down the generational change in frequency as follows:

$$\Delta p = p' - p \\ = (p^2 w_{11} + pq w_{12}) / \overline{w} - p \\ = pq\,[p(w_{11} - w_{12}) + q(w_{12} - w_{22})] / \overline{w} \quad (2)$$

The A_1 allele has increased in frequency if $\Delta p > 0$, and the A_2 allele has increased in frequency if $\Delta p < 0$. This is the result of natural selection at work. No change in gene frequency has occurred if $\Delta p = 0$. allelic equilibrium exists in the system. It should be noted that the absence of natural selection does not necessarily follow from the condition $\Delta p = 0$; rather, $w_{11} = w_{12} = w_{22}$ is required. It is feasible for natural selection to take place without affecting gene frequencies).

Eqns. (1) and (2) explicitly illustrate how genotype-to-genotype fitness differences will result in evolutionary change. This makes it possible for us to investigate the effects of various selective regimes.

First, let's assume that $w_{11} > w_{12} > w_{22}$, i.e., Fitter than the A_2A_2 homozygote, the A_1A_1 homozygote is also fitter than the A_1A_2 heterozygote. Upon examining Eqn. (2), it is evident that Δp needs to be positive (provided that p and q are not zero, in which case $\Delta p = 0$). Thus, until the A_1 allele reaches fixation and the A_2 allele is eradicated from the population, the frequency of the A_1 allele will increase with each generation. Following the A_1 allele's fixation, i.e., If $p = 1$ and $q = 0$, then there won't be any more evolutionary changes because $\Delta p = 0$ if $p = 1$. This makes intuitive sense: the A_1 allele gives carrying organisms a fitness advantage, so its relative frequency in the population will rise over time until it reaches a fixed value.

It is evident that similar reasoning holds true in the scenario where $w_{22} > w_{12} > w_{11}$. According to Eqn. (2), Δp must therefore be negative in order for the A_2 allele to sweep to fixation and eliminate the A_1 allele, provided that neither p nor q is zero.

When the heterozygote outgrows both homozygotes in terms of fitness, an even more intriguing scenario develops. The state in which $w_{12} > w_{11}$ and $w_{12} > w_{22}$. What ought to happen in this case seems obvious intuitively: the population should reach an equilibrium where both alleles are present. Eqn. (2) backs up this feeling. If either allele has gone to fixation, it is obvious that $\Delta p = 0$ (i.e., if either p or q = 0), or, thirdly, if any of the subsequent circumstances holds true:

$$p(w_{11} - w_{12}) + q(w_{12} - w_{22}) = 0$$

Which reduces to:

$$p = p^* = (w_{12} - w_{22}) / (w_{12} - w_{22}) + (w_{12} - w_{11})$$

This condition can only be met if there is heterozygote superiority or inferiority because p must not be negative; it denotes an equilibrium state of the population where both alleles are present. In contrast to the monomorphic equilibria that develop when either allele has gone to fixation, this equilibrium is referred to as polymorphic. It is highly significant that polymorphic equilibrium could occur. It shows us that genetic homogeneity is not always the result of natural selection; occasionally, selection actually protects the genetic diversity present in a population.

A straightforward population-genetic model can be used to answer a great deal of additional questions about natural selection. Population geneticists have brought a level of mathematical rigor to the theory of evolution that would have seemed unthinkable in Darwin's day. For example, by incorporating a parameter that measures the fitness differences between genotypes, we can study the rate of evolutionary change and ask questions like: how many generations will it take for selection to increase the frequency of the A_1 allele from 0.1 to 0.9? If a given deleterious allele is recessive, how long will it take to eliminate it from the population compared to if it were dominant?

Mendelian segregation does not always hold precisely, genotypic fitnesses are unlikely to remain constant across generations, and so on. Selection is rarely the only evolutionary force at work. Creating more realistic evolutionary models, which are more complex and rely on fewer simplifying assumptions, is a major focus of population genetics research. However, the one-locus model exemplifies the fundamentals of population-genetic theory and the ensuing improvements to our understanding of the evolutionary process.

5.3.2. Selection-Mutation Balance

The ultimate cause of genetic variation is mutation, which keeps populations from becoming genetically homogeneous when it otherwise would. A degree of polymorphism will always exist in the population due to recurrent mutation, which ensures that other alleles are present at low frequencies even in cases where one allele at a locus is selectively superior to all others. Exploring the consequences of simultaneous action of selection and mutation has long piqued the interest of population geneticists.

Assuming that the A_1 allele is selectively superior to the A_2 allele in our one-locus, two-allele model, a recurrent mutation from A_1 to A_2 prevents A_1 from spreading to fixation. The frequency of A_1 to A_2 mutations in a generation, i.e., 'u' is the percentage of A_1 alleles that mutate in each generation. (Empirical mutation rate estimates are usually in the range of 10^–6. Since we assume that natural selection has kept the frequency of the A_2 allele in the population extremely low, back mutation from A_2 to A_1 can be disregarded. Remember Eqn. (1) above, which expresses the frequency of the A_1 allele in terms of its frequency in the previous generation. What happens to the gene frequency dynamics under these assumptions? Because a specific fraction (u) of the A_1 alleles will have mutated to A_2, this recurrence equation needs to be changed to.

$$p' = (p^2 w_{11} + pq\, w_{12})\,(1 - u) / \overline{w}$$

to take account of mutation. As before, equilibrium is reached when $p' = p$, i.e., $\Delta p = 0$. The condition for equilibrium is therefore:

$$p = p^* = (p^2 w_{11} + pq\, w_{12})\,(1 - u) / \overline{w} \qquad (3)$$

By adopting a new notation and making certain assumptions about the genotype fitnesses, Eqn. (3) can be effectively simplified. As is frequently the case with deleterious mutants, let us assume that the A_2 allele is fully recessive. This indicates that the fitness of the A_1A_1 and A_1A_2 genotypes is the same. Because of this, genotypic fitnesses can be expressed as follows: $w_{11} = 1$, $w_{12} = 1$, $w_{22} = 1 - s$, where s represents the fitness difference between the A_2A_2 homozygote and the other two genotypes. The selection coefficient against A_2A_2 is denoted by s. Given our assumption of the detrimental nature of the A_2 allele, $s > 0$. Eqn. (3) becomes: when these genotype fitnesses are substituted.

$$p^* = p\,(1 - u) / p^2 + 2pq + q^2(1 - s)$$

Which reduces to:

$$p^* = 1 - (u/s)^{1/2}$$

or equivalently (since $p + q = 1$):

$$q^* = (u/s)^{1/2} \qquad (4)$$

The equilibrium frequency of the A_2 allele, assuming complete recessiveness, is given by Eqn. (4). Keep in mind that q* increases along with u. This makes perfect sense: for a given value of s., the frequency of A_2 that can be maintained at equilibrium will increase with the mutation rate from A_1 to A_2. On the other hand, as s increases, q* decreases. This makes sense as well: for a given

Keyword

Natural Selection – It is the differential survival and reproduction of individuals due to differences in phenotype.

value of u, the equilibrium frequency of A_2 will be lower the stronger the selection is against the A_2A_2 homozygote.

Given that natural selection is continuously eliminating A_2 alleles from the population and mutation is continuously reintroducing them, it is simple to understand why Eqn. (4) is said to represent the selection-mutation balance. Eqn. (4) provides information on the equilibrium frequency of A_2 that will be sustained based on the rate of A_1 to A_2 mutation and the degree of selective disadvantage experienced by the A_2A_2 homozygote. Crucially, Eqn. (4) was developed assuming total recessiveness of the A_2 allele, i.e., that the A_1A_1 homozygote and the A_1A_2 heterozygote share the same phenotype. Comparable equations can be easily derived, nevertheless, in situations in which the A_2 allele is dominant or partially dominant. Because selection is more effective at eliminating dominant or partially dominant forms of A_2 from the population, their equilibrium frequency will be lower if they are dominant or partially dominant than if they are recessive. Recessive harmful alleles can "hide" in heterozygotes and evade selection's eradication process, but dominant alleles cannot.

5.3.3. Random Drift

Random genetic drift, also known as "sampling error," is the term used to describe the random variations in gene frequency that occur in finite populations. To avoid accounting for random fluctuations, the population is frequently assumed to be infinite or extremely large in evolutionary models. However, even though it makes math easier, this assumption is frequently incorrect. In real life, random variables will always be present, especially in small populations. The phrase "random drift" can refer to a variety of stochastic processes that impact gene frequencies in a population, such as chance variations in mating success and survival rates. It can also refer, more narrowly, to the random selection of gametes that result in the next generation of offspring (which occurs when an organism produces many more gametes than will ever fertilize). Here, the wider meaning is employed.

To comprehend the essence of random drift, let's examine a basic illustration. There are just 10 organisms in a population—five of type A and five of type B—and they reproduce asexually, giving rise to offspring of the same type. Assume that both varieties are equally well-suited to their surroundings and that neither is distinctly better than the other. This does not, however, mean that the two varieties will have the same number of children because random variables could be involved. For instance, it is possible that every type B dies accidentally before procreating, in which case there won't be any Bs in the second generation. If this is the case, random drift is the cause of the B type's decline and the A type's subsequent spread. When a change in gene frequency occurs, evolutionists are frequently curious as to whether it is the product of drift, selection, or a mix of the two.

The term "random drift" is a little deceptive. We do not imply that there is no reason for the A type's spread when we state that it is the result of random drift or chance. Theoretically, we could figure out the whole causal chain of events that led to each organism in the population producing the precise number of progeny that it did. We do not deny the existence of a causal story when we attribute the evolutionary change to random drift. Instead, what we mean is that the A type's spread was not caused by its superiority over the B type in terms of adaptability. Stated differently, the A and B types were equally fit because they had the same expected number of offspring, but the A types actually had more offspring than the B types. Evolutionary change results from the nearly universal deviation of actual reproductive output from expected levels in a finite population.

The coin tossing analogy helps explain random drift. Assume that 10 fair coin flips are performed. Since there is a ½ chance of heads on any given toss, 50% of heads should occur in the sequence of 10 tosses. However, the actual likelihood of receiving half heads and half tails is only 242/1024, or roughly 24.6%. Therefore, even if the coin is fair, it is unlikely that we will receive an equal number of heads and tails during a series of 10 tosses; instead, a deviation from expectation is more likely than not. Similarly, even though the A and B types in the previous example are equally fit, there's a good chance that something evolutionary will change. This analogy highlights the significance of population size as well. The percentage of heads would likely be very near to ½ if we tossed the coin 100 times as opposed to ten. Similarly, random drift's influence on gene frequencies decreases with population size; drift has no effect beyond an infinite limit.

Drift makes the population geneticist's job much more difficult. It is evidently impossible to infer the makeup of the population in the second generation from that of the first generation in the example above; the best we can do is infer the probability distribution across all possible compositions. Therefore, no straightforward recurrence relation for gene frequencies of the type given in Eqn. (1) above can be obtained once drift is taken into consideration. Diffusion modeling is a mathematical technique used by population geneticists to analyze the evolutionary effects of drift. A significant consequence of random drift is that it gradually reduces the level of heterozygosity within a population. This is due to the fact that any finite population will eventually drift to homozygosity. This makes sense because gene frequencies of 0 and 1 are known as "absorbing boundaries," which indicate that once they are crossed, there is no turning back (apart from mutation). Thus, an allele will either eventually become fixed in a population or become extinct, with the latter outcome being more likely. In fact, numerical simulations demonstrate that a neutral allele resulting from mutation has an extremely low likelihood of being fixed in a population; the likelihood of fixation decreases with population size.

The different subpopulations of a species will likely diverge genetically as a result of random drift since each will likely experience the chance accumulation of alleles differently, especially if the alleles confer little selective advantage or disadvantage. It is possible that by coincidence, one population will become fixed for allele A_1, while another population will fix itself for allele A_2. This possibility is significant because, if it is disregarded, we might incorrectly assume that, in the first population's environment, the A_1 allele must have been advantageous and, in the second population's environment, the A_2 allele e.g., the genetic differentiation resulted from that selection. While this explanation could be correct, random drift offers a different perspective.

5.3.4. Migration

The last and final factor that can alter a population's genetic makeup is migration—either within or outside of it. Naturally, the genetic composition of the population will change if immigrants differ genetically from the population they are joining. The fact that many species are made up of several unique subpopulations that are mostly isolated from one another but are occasionally linked by migration accounts for the significance of migration in evolution. (Ant colonies are a prime illustration of population subdivision. Gene flow, which results from migration between subpopulations, functions as a kind of "glue," limiting the amount of genetic divergence that subpopulations can experience from one another.

The most basic model for analyzing migration assumes that a population sends out no emigrants but receives a certain number of migrants every generation. Assume that the A_1 allele has a frequency of p in the resident population and a frequency of pm in the population of new immigrants. Every generation, the percentage of migrants entering the population is m (i.e., as a percentage of the total number of residents. Hence, following migration, the population's frequency of the A_1 allele is:

$$p' = (1 - m)\,p + mp_m$$

The change in gene frequency across generations is therefore:

$$\begin{aligned}\Delta p &= p' - p \\ &= -\,m\,(p - p_m)\end{aligned}$$

As a result, if pm > p, migration will increase the frequency of the A_1 allele; if p = pm, it will decrease the frequency of the A_1 allele. Deriving an equation that gives the gene frequency in generation t as a function of its initial frequency and the rate of migration is then a simple task. This is the equation:

$$p_t = p_m + (p_0 - p_m)(1 - m)^t$$

Where; p_0 is the population's initial frequency of the A_1 allele, i.e., prior to the occurrence of any migration. It is evident that equilibrium is reached when pt = pm, i.e., since the expression (1–m)t tends towards zero as *t* grows large., e.g., when the resident population's gene frequency and that of the migrants match.

It is unlikely that migration is the only factor influencing gene frequency at the locus, as this basic model assumes. Therefore, it is important to think about how migration will affect drift, mutation, and selection. Similar to mutation-selection balance, a balance between migration and selection may result in the persistence of a harmful allele in a population. Particularly intriguing is the way that drift and migration interact. As we've seen, drift can cause genetic divergence between a species' various subpopulations. Migration is a homogenizing force that tends to make subpopulations more similar, which runs counter to this trend. According to mathematical models, genetic divergence between a species' subpopulations cannot occur at a rate of migration that is even relatively small. On the grounds that genetic differences between groups, which are necessary for group selection to function, are unlikely to endure in the face of migration, some theorists have used this to argue against the evolutionary significance of group selection.

5.3.5. Non-Random Mating

Remember that the majority of population-genetic analyzes begin with the Hardy-Weinberg law, which was derived assuming random mating. However, deviations from random mating are, in fact, fairly frequent. An arrangement of mating called "positive assortment" occurs when an organism prefers to mate with individuals that are genotypically or phenotypically similar to it. On the other hand, organisms can select "negative assortment" partners—those who are unlike them. Inbreeding, or preferentially mating with relatives, is another kind of deviation from random mating.

While there is a lot of complexity involved in analyzing the effects of non-random mating, certain conclusions are fairly obvious. First and foremost, non-random mating influences genotype frequencies rather than gene frequencies, making it a different kind of evolutionary "force" than selection, mutation, migration, and drift. To understand this, keep in mind that the gene frequency in the pool of successful gametes from which the zygotes are formed is the same as the gene frequency of the population at the zygotic stage. The process by which haploid gametes are "packaged" into diploid zygotes is merely determined by the mating pattern. Gene frequencies will not change if a random mating population suddenly begins to mate in a non-random manner.

Keyword

Population Genetics – It is a subfield of genetics that deals with genetic differences within and among populations and is a part of evolutionary biology.

Second, assortative mating that is positive tends to increase genotypic variance by reducing the fraction of heterozygotes in the population. Examine once more a single locus containing two alleles, A_1 and A_2, with frequencies p and q in a particular population to see this. As a result of the population's initial Hardy-Weinberg equilibrium, 2pq of A_1A_2 heterozygotes exist. If after that the population begins to mate exclusively assortatively, i.e., Since mating only takes place between organisms with the same genotype, the percentage of heterozygotes must inevitably decrease. Only half of the offspring of $A_1A_2 \times A_1A_2$ mating's will be heterozygotic, and there won't be any heterozygotes from $A_1A_1 \times A_1A_1$ or $A_2A_2 \times A_2A_2$ mating's. Therefore, less than 2pq must be the proportion of heterozygotes in the second generation.

What about inbreeding? Inbreeding generally tends to boost population homozygosity, much like positive assortment. This is clearly explained by the fact that relatives typically have higher genotypic similarity than individuals selected at random from the population. "Inbreeding depression" is the term used to describe the detrimental effects of inbreeding on organismic fitness in most species, including humans. This is explained by the fact that harmful alleles frequently have a recessive nature and, as a result, have no phenotypic impact in heterozygotes. Recessive alleles are more likely to be found in homozygotes where their detrimental phenotypic effects become evident when inbreeding lowers the proportion of heterozygotes. The opposite phenomenon, known as "hybrid vigor" due to outbreeding, is frequently employed by breeders of animals.

Remember

Negative assortment will tend to increase the proportion of heterozygotes from what it would be under Hardy-Weinberg equilibrium.

5.3.6. Two-Locus Models and Linkage

Thus far, we have only discussed changes in gene frequency at a single locus, the most basic type of population-genetic analysis. However, there is a limit to how much single-locus models can shed light on the evolutionary process because it is improbable that an organism's fitness will depend on its single-locus genotype in real life. Tracking changes in gene frequency at multiple loci simultaneously is the goal of two-locus (and more broadly, multi-locus) models. These models always attain greater realism than their single-locus counterparts, but they are also more complex.

According to the most basic two-locus model, there are two autosomal loci, A and B, each of which has two alleles, A_1 and A_2, and B_1 and B_2, respectively. Hence, we will designate by the symbols x_1, x_2, x_3, and x_4 the frequencies of the four different haploid gamete types that make up the population: A_1B_1, A_1B_2, A_2B_1, and A_2B_2. (Note that in the case of two loci, we are unable to equate "gamete frequency" with "allelic frequency," unlike in the case of a single locus. The xi are not allele frequencies. Like previously, two gametes fuse to form diploid organisms. By combining every gamete type with every other gamete type, 10 possible diploid genotypes can be identified in the population.

In the single-locus scenario, we observed that the frequencies of the different gamete types and the zygotic genotypes they form in a sizable randomly mated population follow a straightforward relationship. The relationship remains the same in the case of two loci. For instance, the A_1B_1/A_1B_1 genotype frequency will be $(x_1)2$, the A_1B_1/A_2B_1 genotype frequency will be 2×1×3, and so on. Thus, the Hardy-Weinberg law's first part—the genotypic frequencies determined by the square of the array of gametic frequencies—transposes cleanly to the case of two loci. But in the case of two loci, the second feature of the Hardy-Weinberg law—stable genotypic frequencies following one round of random mating—usually does not hold true.

Two-locus population genetics relies heavily on the idea of linkage, or the independency of the two loci. Consider the set of gametes generated by an organism with the A_1B_1/A_2B_2 genotype, i.e., to comprehend linkage e.g., two heterozygotes in one. This set's composition will be {¼ A_1B_1, ¼ A_1B_2, ¼ A_2B_1, ¼ A_2B_2}, if the two loci are unlinked. There will be an equal distribution of all four gamete types. Mendel's first law is assumed to be true at both loci. Because unlinked loci are independent, we cannot infer anything about a gamete's allele at the B locus from its A locus. Perfect linkage represents the other extreme. The set of gametes generated by the A_1B_1/A_2B_2 double heterozygote has the composition {½ A_1B_1, ½ A_2B_2} if the two loci are perfectly linked. This implies that if a gamete receives the A_1 allele at the A locus, it will unavoidably receive the B1 allele at the B locus, and vice versa.

The A and B loci must be physically close to one another on the same chromosome for there to be perfect linkage; as a result, the alleles at the two loci are inherited as a single unit. Unlinked loci are likely to be broken up by recombination because they are either on distinct chromosomes or on the same chromosome but separated by a significant amount. Perfect linkage and total lack of linkage are the two extremes of a continuum where the loci are on the same chromosome. The recombination fraction r, where $0 \le r \le ½$, indicates the degree of linkage. The set of gametes generated by an organism belonging to the A_1B_1/A_2B_2 genotype can be expressed in terms of r as follows:

A_1B_1	$\frac{1}{2}(1-r)$
A_1B_2	$\frac{1}{2}r$
A_2B_1	$\frac{1}{2}r$
A_2B_2	$\frac{1}{2}(1-r)$

It is clear that r = ½ indicates that the loci are unlinked, producing an equal proportion of each of the four gamete types, whereas r = 0 indicates that they are perfectly linked.

Unlike in the one-locus case, in a two-locus model, the gametic (and thus genotypic) frequencies do not have to remain constant across generations, even in the absence of selection, mutation, migration, and drift. Though, in the absence of any evolutionary forces, allelic frequencies will naturally remain constant. Recurrence equations for the gamete frequencies can be written as a function of the recombination fraction plus the frequencies of the previous generation. The ones below are the equations.

$$x_1' = x_1 + r(x_2x_3 - x_1x_4)$$
$$x_2' = x_2 + r(x_2x_3 - x_1x_4)$$
$$x_3' = x_3 + r(x_2x_3 - x_1x_4)$$
$$x_4' = x_4 + r(x_2x_3 - x_1x_4)$$

It is clear from the recurrence equations that genotypic and gametic frequencies will remain constant over generations, i.e., for every i, xi′ = xi, provided that one of two scenarios holds true: (i) r = 0 or (ii) $x_2x_3 - x_1x_4 = 0$. A condition of (i) indicates that the two loci are perfectly linked, functioning as a single locus; a condition of (ii) indicates that the two loci are in a state of "linkage equilibrium," meaning that the alleles at the A and B loci are randomly associated. More specifically, linkage equilibrium indicates that the frequency of the AiBi gamete across the population is equal to the product of the Ai allele and the Biallele.

A significant finding in two-locus theory indicates that, in the event of random mating, the quantity $(x_2x_3 - x_1x_4)$ will diminish with each successive generation until it approaches zero, at which point the frequencies of the genotypes will be in balance. Therefore, over several generations, a population that is initially in linkage disequilibrium will approach linkage equilibrium. The recombination fraction, or value of r, determines the approach rate. Take note of how this differs from the one-locus case, in which genotype frequencies can be brought into equilibrium with a single round of random mating.

Did you get it?

1. **Discuss the non-random mating.**
2. **Explain the two-locus models and linkage.**

5.4. GENETIC-AUTOMATIC PROCESSES

Learning Objectives

- Discuss the effect of isolation and migration on the genetic composition of populations;
- Learn about mutation pressure and selection in natural populations.

The total number of people in a population at any given time is among its most obvious features. Extremely large populations can consist of fewer than a 100 specimens or billions of specimens (microorganisms, algae, and fish populations). In large populations, the distribution and number of specimens are always changing, and this process is happening at different rates. Variations in the population's size have a variety of implications for its evolution. Consequently, the population size influences the frequency of haphazard oscillations in allele concentrations brought on by likely disturbances typical of any small sample size.

It has been demonstrated by Dubinin, Romashov, and Wright that it is theoretically impossible for alleles to remain in stationary concentrations in the absence of mutations, selection, and migration. They demonstrated how errors in the limited selection of gametes under the emergence of the next generation cause concentrations of alleles to undergo rather stochastic changes in any limited population. We refer to these occurrences as genetic-automatic processes.

This procedure is predicated on the idea that, in a small population, the genetic concentration passed down from one generation to the next is not transmitted exactly in the same proportion (as suggested by Hardy's equilibrium law), but rather only as a mathematical expectation. Put another way, there is always a chance that, in a small population with any initial gene concentration, the gene concentration will change in the following generation in either a positive or negative way. The aforementioned claim is crucial to understanding the primary processes that operate within a small population because any alteration in the starting concentration affects the gene's future course. This shift raises the focus to a new level, which serves as the foundation for further growth. Changes in concentration typically occur in many successive stages in different directions, and one of the main causes of their complex course is often their non-direction. In addition, after a number of generations, these sporadic oscillations over a sizable number of generations cause a development that results in drastic changes to the population's structure.

This means that these variations in a small population lead to either the gene's eradication (0% concentration) or its dominance in the population (100% concentration); i.e., the population eventually becomes homozygous in both situations. This change is unavoidable since, in any small population, there exists an ultimate probability of all the concentration changes (because the process is non-directional at each stage) leading from any initial state to homozygous, even though, most of the time, this ultimate probability is very small. But this final possibility is always there, and if it is, it needs to be realized over a very long period of time (Figure 5.2).

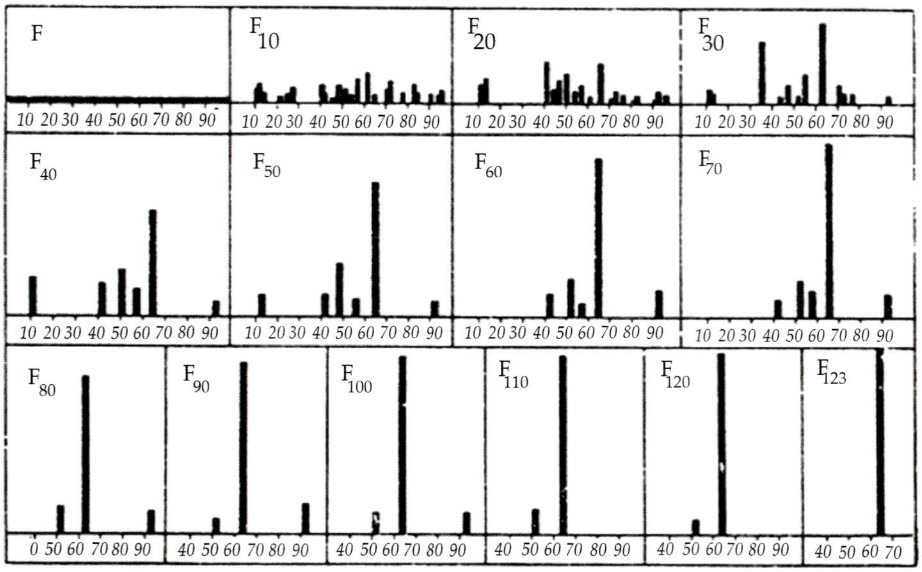

Figure 5.2. *Fluctuations in concentrations of genes based on genetic-automatic processes. Model size: 50 specimens/100 alleles.*

Source: https://www.fao.org/3/b3310e/B3310E104.gif.

Two sets of statistical experiments were conducted by Dubinin and Romashov using models of uniformly multiplying populations with a 100% initial variety of alleles. The first pool used in the experiments had 100 balls in it, each with a different number. 25 balls were scooped out of each drawing, two at a time. One ball represented the perishing allele, while the other represented the multiplying allele. As a result, 25% of the balls were taken out of the pool in each generation, while the remaining 25% doubled. Until every ball in the pool had the same number. The population size in the second series of experiments was different from that in the first. 10 distinct balls made up the first pool.

According to the drawings, the first series' average isogametation period amounted to 268 generations. The outcomes of one of this series' experiments are shown in Figure 5.2. It is evident that only one allele from the original 100% variety still affects the entire population in the 123rd generation. The mean duration of isogametation was equal to 29 generations in the second series of experiments, which was different from the first in that the population size was 10 times smaller. The findings from one of these experiments are displayed in Figure 5.3. In the example given, isogametation occurs over 20 generations. The most intense genetic-automatic processes take place under conditions of uneven multiplication.

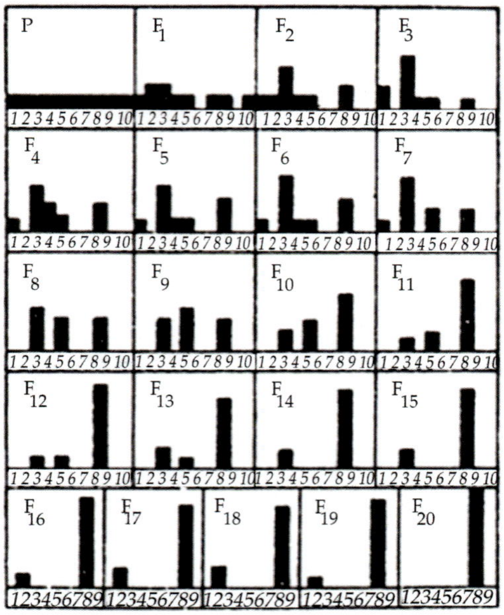

Figure 5.3. *Fluctuations in concentrations of genes based on genetic-automatic processes. Model size: 5 specimens/ 10 alleles.*

Source: https://www.fao.org/3/b3310e/B3310E105.gif.

The following can be used to characterize the general scheme of uneven multiplication under stationary conditions from generation to generation and the size of a population. Every mating pair produces a large number of zygotes, each of which has a very low chance of surviving on its own. Ultimately, only one pair of a parent pair's offspring typically survives to produce offspring in the following generation. However, keep in mind that this is only an estimate, and different pairs actually have different relative importance to the following generation. Some couples have no children at all and couples that have one, two, three, or more children. Fisher and Wright mathematically examined this type of reproduction, which is prevalent in a population with a stationary numerical size. Fisher has demonstrated that, to a certain extent, Poisson's series can be applied in this situation with small gene concentrations ranging from individual genes to concentrations giving rise to homozygous forms.

This series made it possible to calculate the rate at which genes are being replaced and eliminated. Table 5.2 demonstrates that, under this type of multiplication, the population loses 37% of its mutant genes in each generation (compared to 25% in populations that multiply equally). The development and replacement of other genes are also impacted by this value.

Table 5.2. Probability of Elimination and Preservation of Individual Mutations in an Unevenly Multiplying Population with a Stationary Numerical Size Calculated According to Poisson's Series

Generation	Probability		Average Number of Genes in Preserved Mutations
	Elimination	Preservation	
1	0.3679	0.6321	1.58
3	0.6259	0.3741	2.67
7	0.7905	0.2095	4.77
15	0.8873	0.1127	8.87
31	0.9411	0.0589	16.98
63	0.9698	0.0302	33.11
127	0.9847	0.0153	65.36
.	.	.	.
.	.	.	.
.	.	.	.
.	.	.	.
n	$1 - \frac{2}{n}$	$\frac{2}{n}$	$\frac{n}{2}$

Genetic-automatic processes might not matter much in large populations, but they might be crucial in small ones. For example, these processes will be 100 times more intensive in a population of 50 specimens than in a population of 5,000,000 specimens. This example pertains to the population's multiplying component, which is the set of mature specimens from a particular generation that are able to bear children. As an illustration, Figure 5.4 depicts two distinctly distinct population types with equal values for living populations and distinct multiplying components. The rate of isogametation in the two populations differs significantly, and this difference is inversely correlated with the values of the multiplying parts. Despite significant differences in the values of living populations, the rate of isogametation is the same in the four populations depicted in Figure 5.5 because their multiplying parts are also equal.

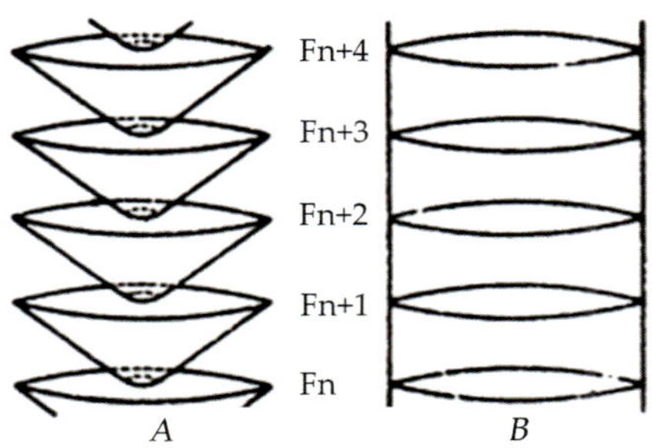

Figure 5.4. *Populations with different ratios of living and propagating parts; under great difference (A); and equality of parts (B).*

Source: https://www.fao.org/3/b3310e/B3310E108.gif.

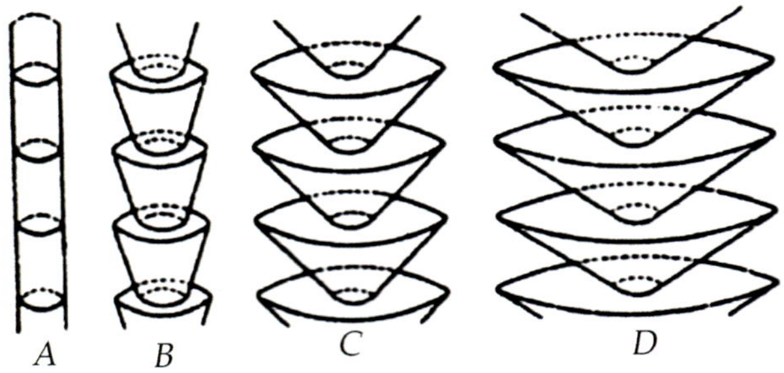

Figure 5.5. *Four types of populations with different ratios of living and propagating parts.*

Source: https://www.fao.org/3/b3310e/B3310E109.gif.

Therefore, it can be said that genetically automatic processes persist for the duration of a population. They occur in populations with a constant number of members. They are particularly intense during times when the population's proliferating portion declines. This time frame is distinguished by a sudden shift in the population's overall genetic makeup. The same procedures are followed as the population grows in size. Even though they are moving very slowly in this instance, they support the continuous differentiation of the species' genetic makeup and dictate its ongoing evolution. The idea that there is a stationary ratio of genes within a population based only on the equilibrium properties of the genotypes surviving throughout any number of generations was thus drastically altered by the study of genetic-automatic processes.

Analyzing the processes involved in a small group of migrants, or even just one fertilized female, populating a new territory is one of the fascinating outcomes of the theory of genetic-automatic processes.

5.4.1. Effect of Isolation and Migration on Genetic Composition of Populations

Distribution in space is one of a population's characteristics, not one of its organisms. Broadly speaking, the distribution area is the smallest geographic area that includes all areas that are habitually inhabited by the specified organisms. Different organisms populate an area in uneven ways. The interactions of the organisms with the animate and inanimate elements of their surroundings typically account for the unevenness of the geographical distribution of specimens.

Different types of isolation and, on the other hand, the migrations associated with population mixing are fundamentally important for the dynamics and history of a species' hereditary pattern of populations. There aren't any species that have cohesive intra-breeding populations in reality. Due to a variety of factors involving both complete and partial isolation, each species within the ecological-geographical area is divided into numerous populations that collectively make up the species.

There is a wide range of isolation factors that determine how a species separates into mostly independent populations. For example, isolation brought on by geographical obstacles like mountains, islands, etc., is extremely common. The isolation of generations over time is interesting. Next, it is pertinent to discuss physiological isolation. The border of fertility, which results in new deviations being more fertile when mating with each other than with the original form, is one way that physiological isolation manifests itself. Multiple groups of specimens that are physiologically isolated from one another can be found in the same habitat.

Ecological isolation brought about by a species' adaptation to the unique conditions of its home station is of great interest. The habitat is not entirely covered by stations, which causes the species to become isolated and disintegrate. The most obvious places for this kind of isolation are where the habitat's boundaries break as the stations encroach on the edge of the nearby ecological-geographical region.

The last and most crucial factor is partial isolation in large, densely populated areas. Because of the specimens' restricted mobility, this isolation happens. Therefore, every species is a complex system made up of reasonably isolated populations that vary in size. The strong differentiation of the genetic material of the species in the many isolates is explained by clearly directed genetic-automatic processes that take place in each of these populations.

However, each species is also impacted by factors that are completely opposite to isolation and that result in population mixing and a particular state of true panmixia. These factors include isolations, all of which have not been covered in this brief report. This mixing occurs at the edges of partially isolated populations, intensifies with more or less significant movements of individual individuals or entire populations, and culminates in periods of migration brought on by sudden increases in population size or environmental changes during crucial stages of the species' life. Panmixia waves, which originated at various points in the species' life cycle, dispersed throughout its habitat and, to some extent, brought previously divided populations together. Naturally, panmixia does not result in the simultaneous and total integration of every population within the species' range. Disintegration of the species into individual populations occurs right after a wave of panmixia (e.g., dispersal of adaptational forms from a certain nidus).

The processes that characterize population mixing are completely different from those that describe isolation. The concentration of distinct genes that were distinctive to individual populations abruptly decreases as a result of integration. Different genes spread uniformly throughout the population, and as a result of the species' shared phenotypic homogeneity, previously noticeable phenotypic differences start to fade. In the end, migrations produce a mean concentration for every mutation across all populations. The following formula affects how quickly the unevenness in gene concentrations is smoothed out, according to calculations:

$$\Delta q = m(q - q_m)$$

Where; 'q' is the frequency of genes in this population; 'm' is the number of migrating specimens divided by the value of the population that accepts them; and 'q_m' is the frequency of genes in the group that is migrating.

The gene frequency distribution with varying m or N is displayed in Figure 5.6, where the gene frequency of immigrants is equal to 0.5. Stationary distributions of frequencies are defined as curves that resemble. Each curve is the function of the density of probabilities of the following series:

$$y - Cq^4Nmq_m - 1\ (1 - q)^{4Nm}(1 - cm)^{-1}$$

Where; the remaining symbols represent the same values as above, and C is the uniformity multiplier. A function like this could display the distribution pattern of the odds of potential outcomes. The probability that the gene frequency will fall within the bounds of the length q is represented by the area between any length of the curve and the corresponding length of the X-axis (axis q). The area under each curve is equal to 1.

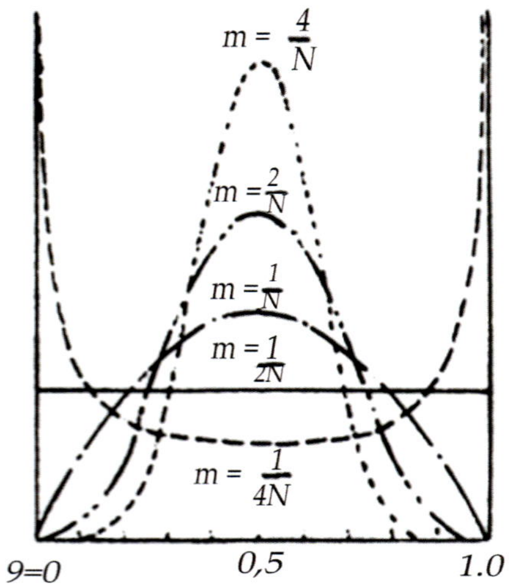

Figure 5.6. *Distribution of gene frequencies among different populations under gene frequency of immigrants: p=q=0.5.*

Source: https://www.fao.org/3/b3310e/B3310E110.gif.

Limits of 0.4 and 0.6 with $m = \frac{1}{4N}$ is far less than with $m = \frac{4}{N}$.

It has been calculated that there is a critical value of migrations above which processes of population mixing and leveling of gene concentrations start to predominate, and below which the distribution of gene concentrations occurs under a particular effect of genetic-automatic processes. This value is equal to $\frac{1}{2N}$ (N stands for the multiplying part of the populations).

It indicates that the distinctive power of genetically automatic processes will diminish due to the leveling influence of migrations and that populations only need to exchange one specimen every generation. One important factor influencing population evolution is the kind and amount of migratory genetic information (gene flow) within and between populations. A one-way gene flow along the linear successions of subpopulations (flow model) is very different from the scenario where members of a group of semi-isolated sub-populations randomly exchange genetic information (island model). In turn, they both diverge from the third scenario, which involves a population of organisms that is uniformly dispersed over a wide region. Despite the possibility that there aren't any semi-discrete groups of specimens in this instance, the mutual remoteness of two distant specimens may make the likelihood of their mating extremely low. The degree of local differentiation within this population is primarily determined by the size of the panmictic concentrations that make up this population. A population with some traits of inbreeding that is split up into semi-isolated subpopulations.

It is important to remember that isolation has detrimental effects when it is sudden, one-sided, and prolonged. Genetically-automatic processes aid in the population's dissolution into groups with distinct, divergent genetic systems that are not crucial for adaptation under such circumstances. However, unrestricted migrations that mix populations of different species can also hinder evolution. In this instance, the very slow rate of evolution of such forms is caused by the emergence of massive united populations with uniform distribution of mutations occurring in accordance with the laws of their low concentrations.

A species' maximum evolutionary flexibility is derived from its differential dynamic system, which arises when population mixing and isolation are optimally correlated. Because of the dynamic condensation of gene concentrations and distinct ecogenotypes, the species' genetic makeup is constantly in tension. The foundation for the evolution of a species is the massive trial and error mechanism that is created for individual populations by the ideal correlation between isolation and migrations.

Keyword

Organism – It is any biological living system that functions as an individual life form

5.4.2. Mutation Pressure and Selection in Natural Populations

When compared to back mutations, the frequency of direct mutations is many times higher. Populations thus take up specific mutational forms as a result of repeated mutation pressure. However, the pattern of the pressure is characterized by factors of direct and reverse mutation that occur in opposite directions due to back mutations. Every gene undergoes two types of mutations: one from a wild, normal allele to a mutational allele, which can then mutate into another allele or return to the normal state. The correlation between the number of normal genes and their mutational alleles in large populations in the absence of selection will be determined by the correlation between the rates of direct and reverse mutation. The concentration of the recessive allele (a) is 1-q, the concentration of the dominant allele (A) is denoted by q, the frequency of mutations from the recessive allele to the dominant (a → A) is represented by V, and the frequency of mutations from the dominant allele to the recessive (A → a) is represented by u.

It is obvious that the distribution of a particular allele within a population affects the quantity of a particular mutation in that population. It can be expressed as the product of the gene's concentration (V(1-q) for reverse mutation and the mutation rate (uq) for direct mutation. The population gradually acquires the gene a if mutations happen in the same direction, that is, if only recessive alleles (A → a) arise from dominants and back mutations do not occur. The allele A will be totally eliminated due to mutation pressure in such circumstances. But these are extremely uncommon cases. Both direct and reverse mutations typically happen.

Regarding the distribution of mutations in nature, the correlation between direct and reverse mutation frequencies is crucial. We can determine the equilibrium concentration when $\Delta q = 0$ (in neutral mutations) by using Δq to represent the rate of changes in the gene's concentration under mutation pressure.

The equilibrium concentration will be reached when the number of normal alleles in the population equals the number of mutational alleles ($q=1-q=0.5$) if direct and back mutations happen at the same rate. The equilibrium concentration will be dependent on the difference in the rate of mutation when there is an unequal rate of direct and back mutations. The equilibrium concentration will be $q = 0.991$ and $1 - q = 0.009$, for example, if the rate of one mutation type is 100 times higher or lower than the rate of the other mutation type. As a result, 99.1% of the population's chromosomes will carry the gene A and 0.9%, the gene A. The population will return to an equilibrium state following subsequent changes in concentrations if the equilibrium concentration is broken.

The rate of these changes (Δq) is determined by the formula:

$$\Delta q = V(1 - q) - uq$$

or by the number of direct mutations $V(1-q)$ minus the number of back mutations (uq).

Neutral mutations are uncommon, though. Selection affects the majority of mutations. Natural selection, which drives out the majority of mutations, is therefore the primary driver of the mutation pressure at the moment of absorbing mutational forms.

ROLE MODEL

MOTOO KIMURA

Motoo Kimura (1924–1994) was a Japanese biologist renowned for his groundbreaking contributions to the field of population genetics and molecular evolution. Born on November 13, 1924, in Okazaki, Japan, Kimura's work revolutionized our understanding of genetic variation, natural selection, and the mechanisms driving evolutionary change.

Kimura's interest in biology was sparked during his undergraduate studies at Kyoto University, where he obtained a degree in agricultural chemistry in 1947. He later pursued graduate studies in genetics under the guidance of Professor Susumu Ohno, a prominent figure in the field. Kimura's doctoral research focused on the genetic basis of coat color variation in mice, laying the foundation for his future investigations into molecular evolution.

In the 1960s, Kimura developed the neutral theory of molecular evolution, which challenged prevailing notions about the role of natural selection in shaping genetic diversity. According to the neutral theory, the majority of genetic mutations that arise within a population are selectively neutral, meaning they neither confer an advantage nor a disadvantage to the organism's survival and reproduction. Instead, genetic variation accumulates in populations primarily through genetic drift, the random fluctuation of allele frequencies over time. This theory revolutionized the field of molecular evolution by providing a null hypothesis against which the effects of natural selection could be tested.

Kimura's seminal work on the neutral theory was published in his influential book "The Neutral Theory of Molecular Evolution" in 1983, where he presented mathematical models and empirical evidence supporting his ideas. His work paved the way for the development of molecular clock methods, which use the rate of molecular evolution to estimate the timing of evolutionary events and divergence between species.

In addition to his contributions to theoretical population genetics, Kimura made significant advances in understanding the molecular mechanisms of DNA replication and genetic recombination. He proposed the "infinite alleles model" to describe the distribution of genetic variation within populations, as well as the "infinite sites model" to analyze nucleotide polymorphism data. Throughout his career, Kimura received numerous accolades and honors for his pioneering research, including the Japan Academy Prize, the Kyoto Prize in Basic Sciences, and election to the National Academy of Sciences in the United States. His work continues to inspire generations of scientists studying the intricate processes of genetic variation and evolutionary change.

Motoo Kimura passed away on November 13, 1994, leaving behind a lasting legacy as one of the most influential figures in the field of molecular evolution. His insights into the neutral theory of molecular evolution revolutionized our understanding of the forces driving genetic diversity and remain foundational to the study of evolutionary biology.

SUMMARY

- The area of genetics known as population genetics focuses on how genetic variation is distributed within populations and how it changes through time.
- Population genetics today is characterized by its emphasis on genetic factors like dominance, epistasis, and the impact of genetic recombination on linkage disequilibrium, setting it apart from phenotypic approaches like adaptive dynamics and evolutionary game theory. This field can be effectively combined with population genomics data to provide a comprehensive understanding of evolutionary processes.
- Though it is the most concrete entity in ecology, the term "individual" is the most ambiguous in the definition. Because individuals in unitary organisms are distinct both genetically and physiologically, identifying them is typically not a problem.
- Mutations, natural selection, non-random mating, genetic drift, and gene flow are some of the factors that can upset the Hardy-Weinberg equilibrium. By introducing new alleles into a population, mutations, for example, upset the equilibrium of allele frequencies.
- Genetic variation interacts with evolutionary forces like natural selection, genetic drift, gene flow, and mutation to produce evolution, the process of population change over time. Heritable variation is subject to natural selection, which benefits people who possess characteristics that improve their chances of surviving and procreating in a particular setting.
- The ultimate cause of genetic variation is mutation, which keeps populations from becoming genetically homogeneous when it otherwise would.

MULTIPLE CHOICE QUESTIONS

1. **The effects of natural selection may be countered by**
    a. gene flow
    b. genetic drift
    c. mutation
    d. inbreeding
2. **Concerning natural selection, evolution is the**
    a. process
    b. outcome
    c. mechanism
    d. purpose
3. **Mating with relatives is called**
    a. inbreeding
    b. outcrossing

c. random mating

d. clines

4. The random loss of alleles in a population is called

a. mutation

b. selection

c. genetic drift

d. gene flow

5. In the Hardy-Weinberg equation, the term 2pq represents the frequency of the

a. dominant homozygotes

b. recessive homozygotes

c. dominant allele

d. heterozygotes

REVIEW QUESTIONS

1. What is population genetics? Describe the properties of populations.
2. Discuss the population research methods.
3. Define the Hardy-Weinberg law.
4. Explain the genetic-automatic processes.
5. Write short notes on:
   a. Random drift;
   b. Migration;
   c. Non-random mating.

Answers to Multiple Questions

1. (a); 2. (b); 3. (a); 4. (c); 5. (d)

REFERENCES

1. Berger, D., & Postma, E. (2014). Biased estimates of diminishing-returns epistasis? Empirical evidence revisited. *Genetics*, *198*(4), 1417–1420.
2. Casillas, S., & Barbadilla, A. (2017). Molecular population genetics. *Genetics*, *205*(3), 1003–1035.
3. Gillespie, J. (2004). *Population Genetics: A Concise Guide* (2nd ed.). Johns Hopkins University Press.
4. Hermisson, J., & Pennings, P. S. (2005). Soft sweeps: Molecular population genetics of adaptation from standing genetic variation. *Genetics*, *169*(4), 2335–2352.

5. Neher, R. A., & Shraiman, B. I. (2011). Genetic draft and quasi-neutrality in large facultatively sexual populations. *Genetics*, *188*(4), 975–996.

6. Orr, H. A. (2010). The population genetics of beneficial mutations. *Philosophical Transactions of the Royal Society B: Biological Sciences*, *365*(1544), 1195–1201.

7. Rajon, E., & Masel, J. (2011). Evolution of molecular error rates and the consequences for evolvability. *Proceedings of the National Academy of Sciences*, *108*(3), 1082–1087.

CHAPTER 6

Human Genetic Variation

LEARNING OBJECTIVES

After studying this chapter, you will be able to:

- Give an overview human genetic variation
- Discuss the alteration in human genetic variation
- Focus on existence of genetic variation among humans
- Explain the significance of human genetic variation
- Discuss the human genetic variation and medicine
- Give the overview of genetics, ethics, and society

INTRODUCTORY EXAMPLE

An Introduction to Human Genetic Variation

The term "human genetic variation" describes the variations in DNA sequences found in different members of the human population. These differences result from recombination, mutations, and other genetic processes that take place over many generations. Despite the fact that 99.9% of human DNA is shared, the 0.1 percent that differs between individuals accounts for notable variations in characteristics like physical attributes, disease susceptibility, and reactions to environmental stimuli. This variation allows populations to flourish in a variety of environments, which is crucial for human evolution and adaptability.

Single nucleotide polymorphisms (SNPs), which are tiny variations in the DNA sequence, as well as copy number variations (CNVs) and structural variants, which entail larger-scale changes, are the most prevalent types of genetic variation. These variations may have an impact on an individual's drug metabolism, susceptibility to infections, and development of diseases like diabetes or cancer. Moreover, genetic diversity is essential to personalized medicine because it makes it possible to customize a patient's course of treatment based on their unique genetic profile.

Despite their great similarity, human populations exhibit genetic variation patterns influenced by natural selection, migration, and historical events. For instance, differences in skin tone have developed as a result of localized adaptations to differing UV radiation levels. Furthermore, some genetic mutations confer benefits; for example, populations with a history of dairy farming may develop lactose tolerance, or regions with a high malaria prevalence may develop resistance to the disease.

Although there are many advantages to genetic variation, there are drawbacks as well, such as moral questions regarding discrimination, privacy, and unequal access to genetic technologies. Gaining a deeper understanding of human diversity and advancing medical research and healthcare depend on our ability to comprehend human genetic variation.

UNIT INTRODUCTION

The genetic variations within and between populations are known as human genetic variation. Polymorphism is the term used to describe the possibility of multiple variants (alleles) of a single gene in the human population. Humans are not genetically alike. Gene copy-number variation and developmental mutations are the rare causes of genetic differences, even in monozygotic twins (who develop from a single zygote). Techniques like genetic fingerprinting rely on individual differences, even amongst closely related individuals.

The human genome is made up of slightly less than 17,000 base pairs (bp) of DNA in cellular mitochondria and approximately 3.2 billion base pairs (bp) total across 46 chromosomes. According to estimates from 2015, the average difference between a person's genome and the reference genome was 20 million base pairs, or 0.6 percent of the total genome. 324 million distinct variations from human genomes that have been sequenced have been identified as of 2017.

In contrast, humans belong to a genetically homogeneous species. A tiny fraction (~15%) of the variability in the human genome is explained by the few genetic variants that are more common in particular geographic areas or in individuals descended from those regions. Most variation is found among the individuals that make up each human population. In contrast, the DNA sequence diversity of rhesus macaques is 2.5 times higher than that of humans. Depending on the macromolecules under study, these rates vary. When looking at nuclear DNA, chimpanzees have more genetic variation than humans; however, when looking at proteins, humans have more genetic variation.

The striking homogeneity of human populations worldwide, the lack of discontinuities in genetic distances between human populations, and the absence of distinct branches in the human species all suggest that there is no scientific basis for identifying racial or subspecies distinctions in humans and that variation within populations is typically greater than that between populations for the majority of traits. In spite of this, contemporary genetic research has discovered significant average genetic variations among human populations concerning characteristics like skin tone, body size, the ability to digest lactose and starch, adaptation to high altitudes, drug response, taste receptors, and susceptibility to specific illnesses. According to the Out of Africa theory of human origins, the greatest diversity is found within and among African populations, and it gradually decreases with increasing distance from the continent.

There are evolutionary and medical implications to the study of human genetic variation. It can assist researchers in reconstructing and comprehending historical human migration patterns. The study of human genetic variation may be crucial to medicine because certain population groups have higher frequencies of certain disease-causing alleles. For example, the mutation that causes sickle-cell anemia is more common in people descended from populations in sub-Saharan Africa, South Europe, Arabia, and India. This is because these populations have experienced evolutionary pressure from malaria-carrying mosquitos.

6.1. OVERVIEW HUMAN GENETIC VARIATION

Learning Objectives

- What is the human genome?
- How do peoples' genomes vary?
- What are the different types of genomic variants?
- How are genomic variants detected?
- What is a reference genome sequence?

The scientific study of inherited variation is called genetics. Thus, the scientific study of inherited human variation is known as human genetics. One reason to study human genetics is the simple desire to learn more about ourselves. Human genetics is a field of study that focuses on Homo sapiens, which is widely regarded as the most fascinating species on the planet. However, our interest in human genetics does not end with species boundaries because, in the same way, that the study of variation in other species informs our understanding of our own, what we learn about human genetic variation and its sources and transmission inevitably contributes to our understanding of genetics in general.

The usefulness of studying human genetics for promoting human welfare is a second motivation. Thus, rather than being a fundamental science, human genetics is more of an applied science. The identification and characterization of the genetic component of numerous human diseases is one advantage of researching human genetic variation. Given our expanding knowledge of the role that genes play in the onset of diseases like diabetes, cancer, and heart disease, this is an increasingly compelling incentive. In actuality, society is willing to pay a substantial sum of money for research in this field and has done so in the past, mainly because it believes that the field has a great deal of potential to improve human health. The realization of this perception in discoveries in the last two decades has resulted in a notable surge in the number of individuals and institutions dedicated to the field of human genetics.

The first and second reasons for researching human genetics are connected. Basic research has been strongly supported by the goal of creating medical procedures that can lessen the suffering brought on by human disease. Investigations into specific medical conditions have led to the discovery and description of numerous fundamental biological phenomena. One well-known example is the understanding of human sex chromosomes that resulted from researching individuals with abnormal sex chromosomes. A more recent example is our rapidly developing knowledge of the mechanisms governing cell division and growth, knowledge primarily derived from the study of genes that, when mutated, raise the risk of cancer.

In a similar vein, basic research findings influence and motivate investigations into human illness. For instance, the rapid advancement of recombinant DNA techniques (Figure 6.1) revolutionized the field of human genetics by enabling researchers to examine the precise composition and roles of individual human genes and to modify them in a range of previously unthinkable ways.

Keyword

Human Genetics – It is the study of inheritance as it occurs in human beings.

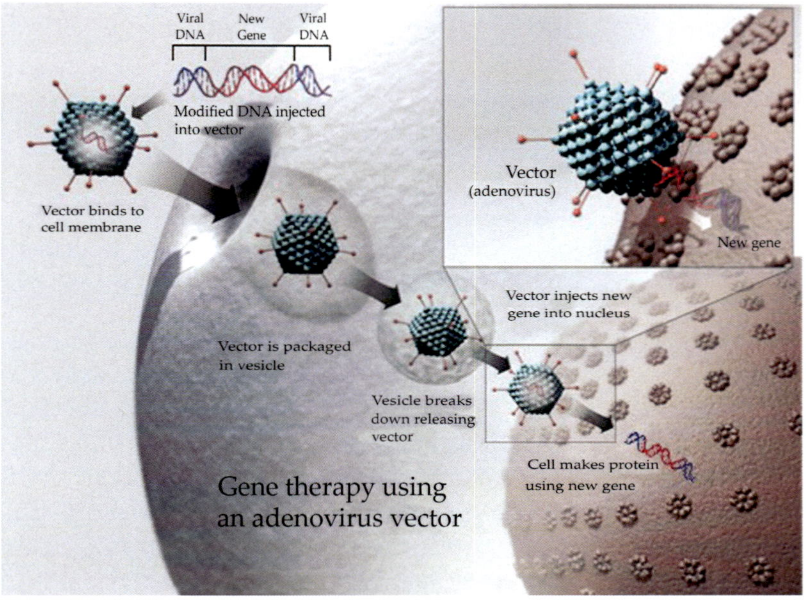

Figure 6.1. *Recombinant techniques have transformed the study of human genetics.*

Source: Public Domain, https://commons.wikimedia.org/w/index.php?curid=488192.

The fact that studying human genetics provides us with an effective tool for comprehending and characterizing human evolution is a third reason for doing so. Scholars interested in tracing the evolutionary history of humans had to rely exclusively on data from physical anthropology, which included information about skin color, body build, and facial traits. However, a multitude of genetic data, including molecular data, is now available for researchers to use in their research.

6.1.1. What Is the Human Genome?

The entire collection of DNA instructions present in every cell is called a genome. The four distinct chemicals that make up DNA are referred to as nucleotides, or bases, and are denoted by different letters: adenine (A), thymine (T), cytosine (C), and guanine (G). These letters' sequence (i.e., the information telling each cell what to do and when to do it is encoded in the DNA sequence. The DNA that makes up the genome is arranged into units called chromosomes (Figure 6.2).

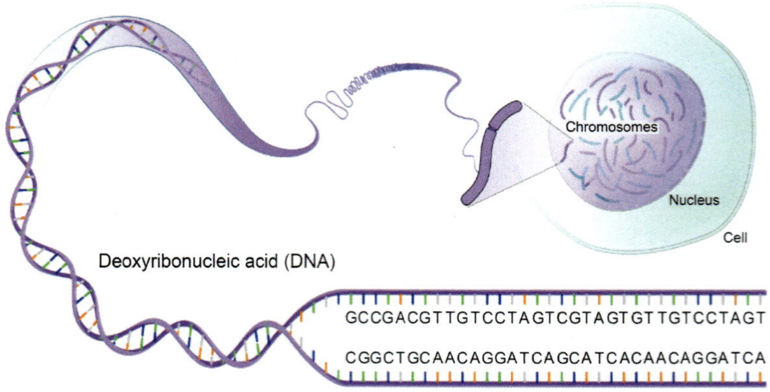

Figure 6.2. *A genome consists of all the DNA in a cell, which is packed into chromosomes that reside in the nucleus of cells.*

Source: https://www.genome.gov/sites/default/files/inline-images/Genomic%20variation_CHROM-DNA_SEQ.png.

The human genome is composed of approximately 3 billion nucleotides per copy, which are split up among 23 chromosomes. The human genome is found in two copies in the majority of human cells, one copy coming from each parent. "Diploid" cells are those that have two copies of every chromosome. Although most mammals are diploid, some species only have one set of each chromosome or more than two sets (Figure 6.3).

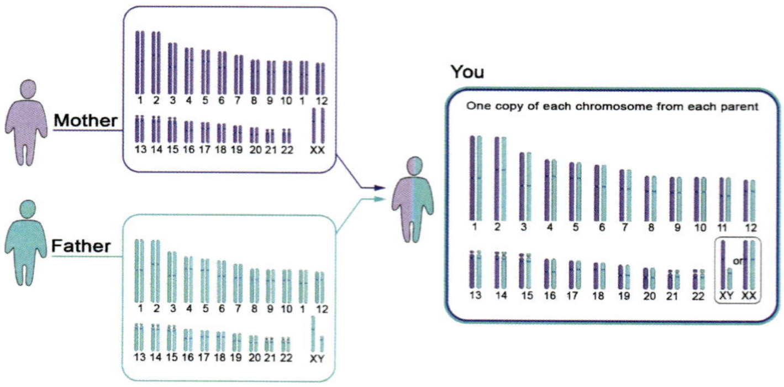

Figure 6.3. *A person's genome consists of 23 pairs of chromosomes. One copy of each chromosome is inherited from each parent. For example, chromosomes from the depicted mother are labeled in purple, and chromosomes from the depicted father are labeled in teal. The X and Y chromosomes are known as sex chromosomes. Although there are exceptions, biological females have two X chromosomes (denoted XX) and biological males have one X chromosome and one Y chromosome (denoted XY).*

Source: https://www.genome.gov/sites/default/files/inline-images/Genomic%20variation_Diploid.png

6.1.2. How Do Peoples' Genomes Vary?

Genomes share far more similarities than differences between individuals. The claim that the genomes of any two individuals are approximately 99.9% identical is made often. This percentage is based on the discovery that, roughly every 1,300 nucleotides, there is a single nucleotide difference between the genomes of two individuals.

But since it only takes into consideration variations in a single nucleotide, this oversimplifies the situation. In actuality, the average genome of two individuals is approximately 99.6% identical and 0.4% different. The latter percentage represents differences involving multiple nucleotides as well as differences involving single nucleotides.

Genomic variants are the variations in human genomes. One aspect of individuality is the collection of genetic variants they possess. A person's uniqueness is also influenced by other elements, including their diet, lifestyle, surroundings, and social background. A tiny fraction of genomic variants affect how well a person's genome functions, but the majority of variants have no effect at all. For instance, certain genetic variations affect physical traits like height and eye color, while others affect medical conditions or a person's reaction to specific drugs.

6.1.3. What Are the Different Types of Genomic Variants?

6.1.3.1. There Are Multiple Types of Genomic Variants

Single-nucleotide variants (SNVs) are the smallest genomic variants. A difference in a single nucleotide (or letter) is reflected in each SNV. The DNA letter at that genomic location for a particular SNV may be a T in one individual but a C in another. The most prevalent kind of genomic variation is called a SNV. Single-nucleotide polymorphisms (SNPs; pronounced "snip") are a subtype of SNVs. A SNV needs to be present in at least 1% of the human population in order to be classified as an SNP. Consequently, SNV is a broader term that encompasses both uncommon and relatively common single-nucleotide differences (SNPs, for example). Regardless of their relative frequency, we will refer to all single-nucleotide differences as SNVs for the sake of simplicity (Figure 6.4).

"Indels," or insertions and deletions, are another class of small genomic variants. The majority of insertion/deletion variants involve fewer than 50 nucleotides and represent either extra or absent DNA nucleotides in the genome. While less common than SNVs, insertion/deletion variants can occasionally have a greater influence on health and disease (e.g., by interfering with a gene that produces a crucial protein).

One of the most common types of insertion/deletion variants are tandem repeats (also known as microsatellites). Tandem repeats are short, highly variable segments of nucleotides that are repeated several times. The frequency with which these brief nucleotide stretches are repeated can differ amongst chromosomes, ranging from a few to hundreds of times. Tandem repeats have been employed in the past to map the human genome and profile DNA for forensic purposes (Figure 6.5).

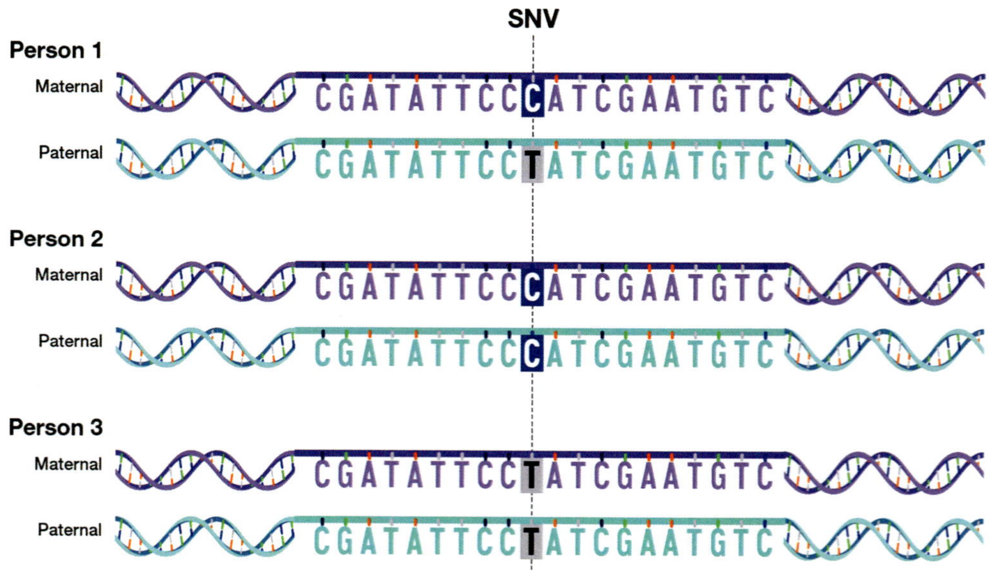

Figure 6.4. *Single-nucleotide variants (SNVs) are differences of one nucleotide at a specific location in the genome. An individual may have different nucleotides at such a location on each chromosome (getting a different one from each parent), such as with Person 1. An individual may also have the same nucleotide at such a location on both chromosomes, such as with Person 2 and Person 3.*

Source: https://www.genome.gov/sites/default/files/inline-images/Genomic%20variation_SNV.png.

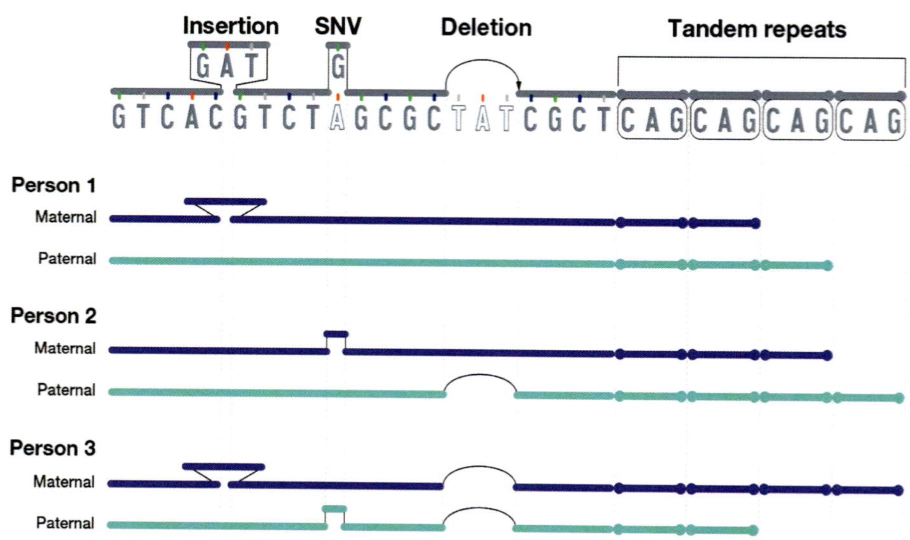

Figure 6.5. *Each person has a collection of different genomic variants. For example, Person 1 has an insertion variant; Person 2 has a SNV and deletion variant; and Person 3 has an insertion, SNV, and deletion variant. All three people have different tandem repeats. Different variants can be inherited from different parents, either the maternal genome (purple) or paternal genome (teal).*

Source: https://www.genome.gov/sites/default/files/inline-images/Genomic%20variation_INDELS%20DEL%20SNV%20TANDEM.png.

Larger chromosomal regions are likewise subject to genomic variation, which goes beyond short nucleotide segments. These extensive genomic alterations, known as structural variants, involve thousands of nucleotides that have been added, removed, flipped, or transferred from one region of the genome to another, in addition to at least 50 nucleotides. More than 50 nucleotides in a tandem repeat are regarded as a structural variant; in fact, nearly half of all structural variants found in human genomes are composed of such large tandem repeats. A structural variant is referred to as a copy-number variant (CNV) when it shows variations in the total number of nucleotides involved. Keep in mind that CNVs are different from other structural variations like translocations and inversions because the latter kinds frequently don't involve a change in the total number of nucleotides (Figure 6.6).

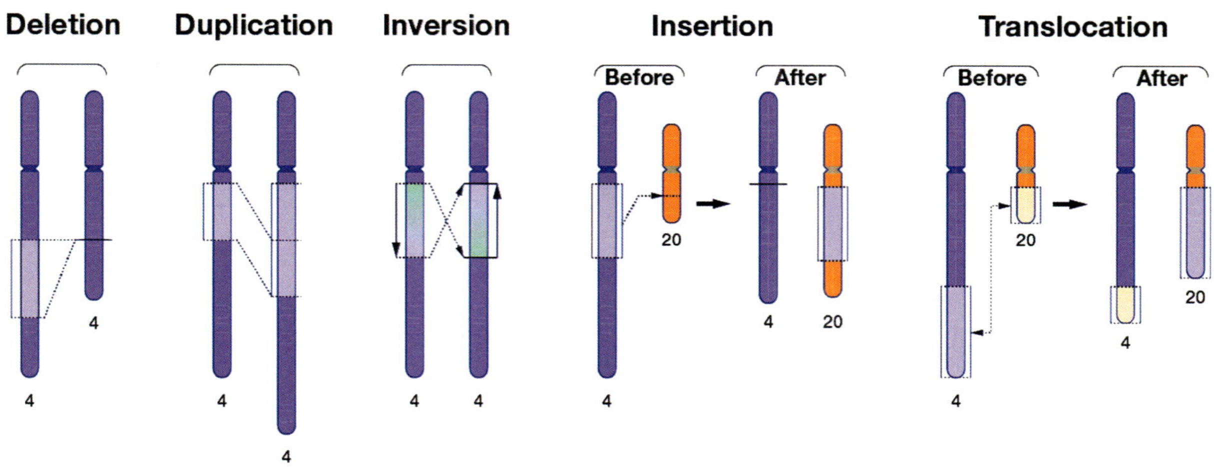

Figure 6.6. *Larger structural variants are present among human genomes. For example, human chromosomes can have missing segments (deletion variants), duplicated segments (duplication variants), inverted segments (inversion variants), added segments (insertion variants) and segments transferred from other chromosomes (translocation variants).*

Source: https://www.genome.gov/sites/default/files/inline-images/Genomic%20variation_Structral%20variants.png.

6.1.4. How Are Genomic Variants Detected?

Certain techniques are intended to identify only known genetic variations. Researchers can use a specific DNA test to find out whether a person has a genomic variant that could be relevant to their health or treatment, for instance, if it might be important to know if they have inherited it. Using alternative techniques, scientists can examine an individual's DNA to check for a wide range of known genomic variants. For instance, a test known as a microarray can simultaneously identify hundreds of thousands of SNVs.

Genome sequencing is a more thorough method of identifying genomic variants, including ones that may not be known as of yet. Human genomes can now be sequenced using a variety of techniques, all of which only need a tiny sample of cheek, hair, or blood cells from which DNA can be extracted. A human genome can now be sequenced for less than $1,000, which is over a million times less expensive than it was a few decades ago.

Both copies of each pair of chromosomes—one from each parent—are sequenced simultaneously when a person's genome is sequenced. Most commonplace circumstances (e.g., the parent of origin of each discovered genomic variant is not always easily ascertained (especially when a genome is sequenced for medical purposes). This is due to the fact that the human genome is not sequenced one chromosome at a time, but rather the genome is broken up into smaller pieces and then the pieces are put back together, much like a jigsaw puzzle. That procedure does not reveal the parental origin of any given genomic variant without further analysis. To sequence the human genome and find genetic variants, a "reference" sequence is also essential.

6.1.5. What Is a Reference Genome Sequence?

A reference sequence must typically already exist in order to detect variations in an individual's genome sequence. A reference sequence of the human genome is a recognized, reliable, and approved sequence of the genome (i.e., 23 human chromosomes, one sequence for each.

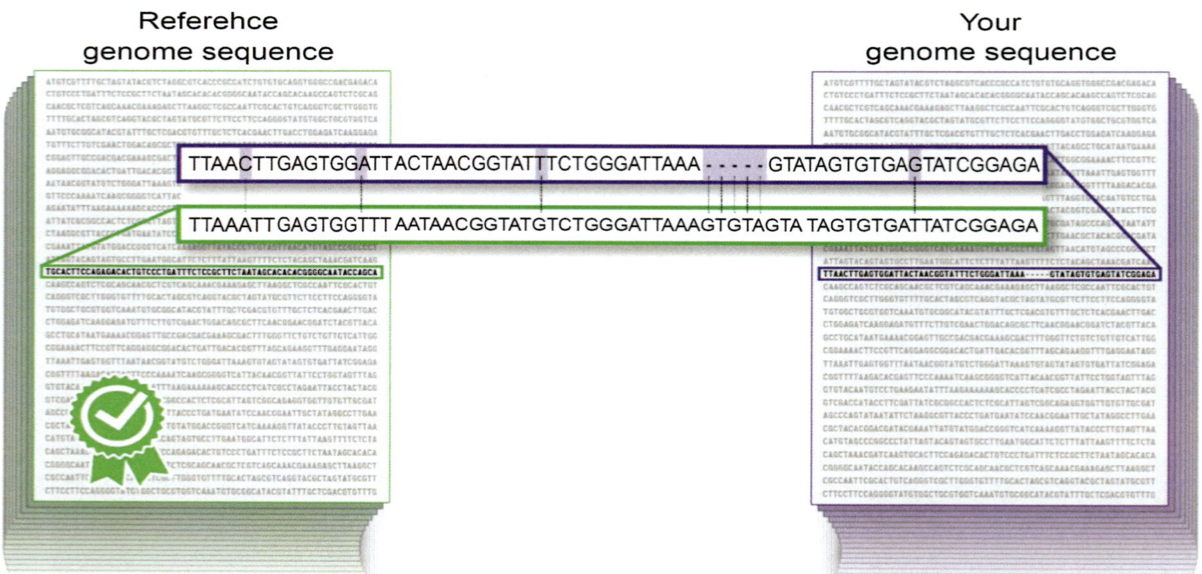

Figure 6.7. *A reference genome sequence (green) is used for comparing individual human genome sequences (purple) to find genomic variants. Specifically, the side-by-side comparison of a newly generated human genome sequence with a reference human genome sequence allows for the detection of genomic variants in the former. Although only one generated human genome sequence is shown for the purposes of simplification, in reality, each person actually has two genome sequences (one inherited from each parent), both of which would be compared to the reference genome sequence.*

Source: https://www.genome.gov/sites/default/files/inline-images/Genomic%20variation_your%20Genome%20vs%20Reference2.png.

Instead of being an exact sequence from a single person, a reference human genome sequence is assembled from several individuals. An essential characteristic of a reference human genome sequence is that each position in the human genome is represented by a single assigned nucleotide. Regarding this, the sequence of a reference human genome only reflects the sequence of one copy of each chromosome, while the sequence of an individual's genome comprises the sequences of both copies of each chromosome (i.e., an individual is diploid) (Figure 6.7).

Genome variation within the human population is extremely high. The purpose of a given reference human genome sequence is to be a useful tool for data analysis, not to capture or represent that diversity. To be more precise, every newly generated genome sequence from an individual is compared directly to a reference human genome sequence, making it possible to identify any variations (differences) between the individual's genome and the reference genome. To put it briefly, all genomic variants in an individual's genome can be cataloged by comparing their genome sequence to a reference genome sequence.

6.1.6. What Is the Inventory of Genomic Variants in a Typical Human Genome Sequence?

What does a typical human genome sequence look like with respect to the variants that it contains? On average, compared to a reference human genome, a person's ~6 billion-nucleotide genome sequence will have:

- ~5,000,000 SNVs that involve ~5,000,000 nucleotides
- ~600,000 insertion/deletion variants that involve ~2,000,000 nucleotides
- ~25,000 structural variants that involve >20,000,000 nucleotides.

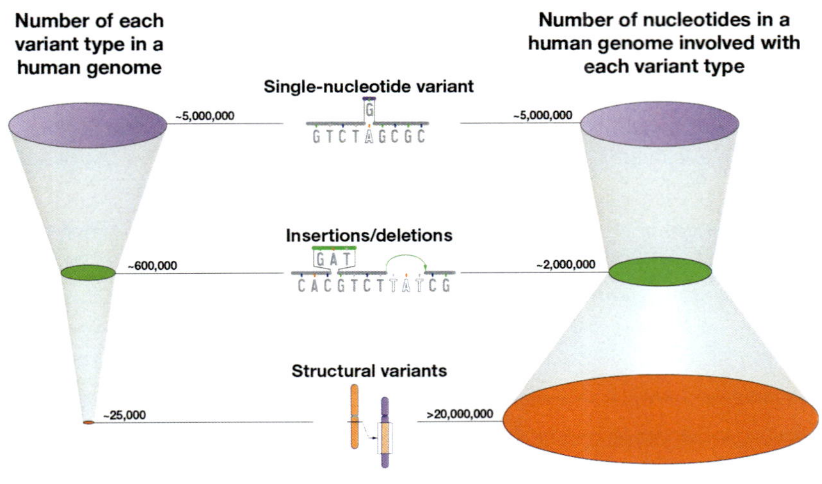

Figure 6.8. *On average, a person's genome sequence contains the total numbers of each of the three major types of genomic variants indicated on the left. Together, those genomic variants involve the total number of nucleotides indicated on the right.*

Source: https://www.genome.gov/sites/default/files/inline-images/Genomic%20variation_Space%20filled%20comparison.png.

Note: Each SNV only involves a single nucleotide, whereas each insertion/deletion and structural variant involves more than a single nucleotide. As a result, each type of genomic variant differs with respect to its overall contribution to genomic variation in terms of the number of nucleotides involved (e.g., a given structural variant contributes substantially more than a given SNV). Overall, insertions/deletions (in green) are less common than SNVs (purple) and involve fewer nucleotides, whereas structural variants (in orange) are less common than either SNVs or insertion/deletions but involve the greatest number of nucleotides.

These figures are current estimates of averages and will probably vary as more complete genomes of humans are sequenced. Furthermore, the amount by which the various kinds of genomic variants explain the variations in genomes can be computed using a variety of methods. In actuality, figuring out how many nucleotides are involved in all of the structural variants in each human genome is a challenging task that is still being researched (Figure 6.8). This indicates that, of the approximately 6,000,000,000 nucleotides in each person's genome, the full set of genomic variants involves, on average, ~27,000,000 nucleotides. Together, those approximately 27,000,000 nucleotides represent approximately 0.4 percent of the individual's entire genome and each nucleotide reflects a variation at those specific locations in the DNA. Put another way, a typical person's genome sequence is approximately 99.6 identical to (or ~0.4 percent different from) a reference human genome sequence (or even another person's genome sequence) when taking into account the entire inventory of genomic variants (Figure 6.9).

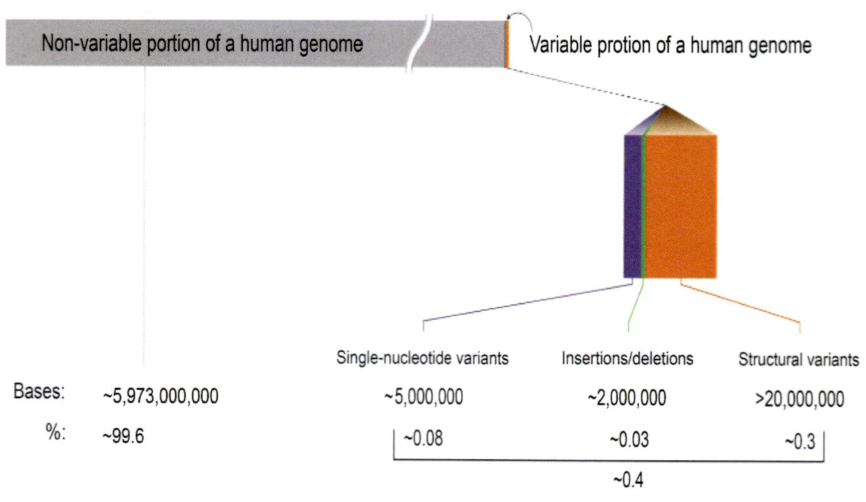

Figure 6.9. *On average, a human genome sequence (consisting of ~6,000,000,000 nucleotides) is ~99.6% identical — or non-variable — compared to a reference human genome sequence. The variable portion of a human genome sequence totals ~0.4%, which consists of the indicated contributions from each of the three major types of genomic variants. These same general numbers are also found when directly comparing two peoples' genome sequences (as opposed to comparing to a reference human genome sequence).*

Source: https://www.genome.gov/sites/default/files/inline-images/Genomic_variation_barchart_VariableNONvariable.png.

6.1.6.1. What Is a Pangenome Reference Sequence and How Is It Used?

Genomics variants in every newly sequenced human genome can be found thanks to reference genome sequences. However, there is a lack of genomic diversity in the human population represented by the reference human genome sequences that are currently available, and this lack of diversity can introduce biases when analyzing the genome sequences of some individuals.

To address this deficiency, researchers are working to generate a more complete set of reference human genome sequences that better reflect all of humanity. The combination of several individuals' genome sequences to better represent the genomic diversity of a species is called a "pangenome.". To better identify and characterize genomic variants, including those that play significant roles in health and disease, a more comprehensive comparison of genome sequences from individuals worldwide will be made possible by the creation of the human pangenome reference sequence.

Keyword

Genomic medicine is a medical field that uses a person's DNA to personalize their healthcare.

6.1.6.2. Why Does Genome Variation Matter?

Genomic variation promotes biodiversity expansion and drives evolution. This holds true for living things such as plants, animals, and humans. Natural selection, the process through which organisms evolve to adapt to changing environments, depends on this kind of diversity to maintain healthy populations. Variations in human genomes are also very important for medicine. The majority of genomic variants have no effect on human health. Certain diseases (like cystic fibrosis and sickle cell disease) are caused directly by genetic variants in certain cases. When it comes to conditions like diabetes and hypertension, where a genetic variant may raise an individual's overall risk of developing the illness, the effects of genomic variants are more subdued. 'Genomic medicine' is the use of patient genetic variant data to guide medical decisions, and it is a field that is rapidly expanding among healthcare professionals.

Did you get it?

1. **What is the inventory of genomic variants in a typical human genome sequence?**
2. **What is a pangenome reference sequence and how is it used?**

6.2. ALTERATION IN HUMAN GENETIC VARIATION

Learning Objectives

- What are genetic alterations?
- Genetic Alterations, Mutations, and Polymorphisms

In the past, two research strategies were crucial in assisting scientists in comprehending the biological underpinnings of heredity. Using the first of these methods, transmission genetics, theories regarding the mechanisms of inheritance were developed by crossing organisms and examining the traits of the progeny. This work showed that heredity appears to follow a few clear and relatively simple rules, at least in some organisms.

The second strategy involved examining the mechanisms and procedures of cellular reproduction through the use of cytologic techniques. The more conceptual understanding of inheritance that emerged as a result of transmission genetics was made possible by this method. By the turn of the 20th Century, cytologists had established the following: that gametes are the means by which genetic information is passed down from one generation to the next; that the nucleus and chromosomes play a central role in heredity; and that heredity results from the genetic continuity of cells through cell division.

Even though they were significant, the methods of cytology and transmission genetics were insufficient to enable scientists to comprehend human genetic variation at the level of detail that is now achievable. The primary benefit that modern molecular techniques provide is the ability to directly study DNA. Prior to the creation of these methods, researchers looking at genetic variation in humans had to extrapolate molecular differences from the phenotypes caused by mutant genes. Furthermore, only a limited number of families could have their genes studied because the majority of single-gene disorders are linked to relatively uncommon genes. Numerous characteristics linked to these genes are likewise recessive, meaning that individuals with heterozygous genotypes would not exhibit them. In contrast to scientists studying other animals, human geneticists are prohibited from conducting "at-will" experiments on human subjects due to ethical concerns. Furthermore, human generations span between 20 and 40 years, making them far too slow to be effective in traditional breeding experiments. Due to all of these restrictions, identifying and researching human genes was time-consuming and slow.

But over the past 50 years, a plethora of molecular techniques have made direct study of human DNA a reality, starting with the discovery of DNA's structure and picking up speed with the introduction of recombinant DNA techniques in the mid-1970s. These methods include, among others, the following: hybridization techniques like fluorescence in situ hybridization, which enable scientists to compare DNA samples from various sources and identify specific base sequences within samples; amplification techniques like polymerase chain reaction (PCR), which enable researchers to cut and rejoin DNA molecules in highly specific and predictable ways; and automated sequencing techniques, which today are enabling workers to sequence the human genome at an unprecedented rate.

Remember

It is a cutting-edge instrument that can be used to detect gene mutations or measure the expression of tens of thousands of genes in a single experiment.

More potent methods are just around the corner, and experts predict they will have a significant influence on clinical genetics and research in the future. One such method is DNA chip technology, also known as DNA microarray technology.

The chip's intended use in this technology is to identify changes in a specific gene. The tiny glass plate that makes up the DNA microchip is covered in plastic. It is produced by a method that is comparable to that of producing computer microchips. Its surface is covered in synthetic single-stranded DNA sequences that are exact replicas of the gene's normal sequence as well as every conceivable gene mutation. A scientist first takes a sample of DNA from the subject's blood, along with a sample of DNA that does not have the gene mutation, to find out if the subject has the mutation. The scientist labels the DNA fragments with fluorescent dyes, labeling the individual's DNA with red dye and the normal DNA with green dye, after denaturing, or separating, the samples into single strands and cutting them into smaller, more manageable fragments. The synthetic DNA on the chip is open to hybridization, or binding, between the two sets of labeled DNA. Both DNA samples will hybridize similarly to the chip if the person does not have a gene mutation, making the chip appear uniformly yellow. However, the chip's mutant sequence will manifest if the individual does carry a mutation.

Researchers can now simultaneously examine thousands of genes' activities in various cell types thanks to DNA microarray technology, which is an advancement that will aid in understanding the intricate functional relationships that exist between individual genes. In this kind of analysis, fluorescently labeled mRNA molecules from a specific cell type are allowed to hybridize with small segments of DNA spanning hundreds or thousands of genes on a single microscope slide. Scientists can measure the fluorescence of each spot on the slide to ascertain the relative activity of different genes in that type of cell. Robust fluorescence suggests that a high number of mRNA molecules have hybridized to the gene, indicating high gene activity in that particular cell type. On the other hand, the absence of fluorescence signifies that the gene is dormant in that particular cell type and that none of the mRNA molecules in the cell hybridized with it.

Scientists anticipate that these relatively new technologies will eventually find extensive clinical use, even though their current primary use is in research. For instance, the time and cost of genetic testing may be greatly decreased with the use of DNA chip technology. One day, this technology and others similar to it might help determine a person's risk of numerous hereditary cancer types as well as other common illnesses like diabetes and heart disease. Similarly, scientists might eventually be able to categorize human cancers according to the gene activity patterns in the tumor cells, and they could then create treatment plans that are specifically tailored to each individual type of cancer.

6.2.1. What Are Genetic Alterations?

Any modification to the DNA sequence has the potential to change the genetic code, which in turn could change how the protein it encodes is synthesized. A change in one nucleotide (C for A) alters the composition of the protein, potentially impairing its function. For instance, the genetic code table shows that the amino acid glutamine is translated from the CAA codon, while lysine is translated from the AAA codon. However, if the alteration is to UAA, protein synthesis is halted because this stop codon prevents the production of glutamine.

Keyword

Cells are the basic building blocks of life, making up all living organisms and the tissues of the body

As a result, the location of a genetic change will determine its clinical significance, i.e., whether the change occurs in the coding region (exome) or not, and whether it has a significant impact on how proteins are synthesized and, consequently, function in the body.

6.2.1.1. Why Do Genetic Alterations Occur?

There are two main causes for genetic alterations:

- External, environmental factors.
- Internal, genetic factors.

In the human body, almost all **cells** are constantly being replaced. The cells split into two daughter cells to accomplish this. This division process is prone to errors that result in genetic changes. Among many other external factors, smoking and sun radiation raise the chance of these kinds of errors occurring. Since these mutations only impact the offending cell and are not inherited by the progeny, we refer to them as somatic mutations.

On the other hand, genetic changes may also exist from birth. Since all of the cells that make up the "new" human being are descended from the "original" cell, any genetic errors in the egg or sperm cell will be passed on to the zygote and will thus manifest in all of its cells. Even if the change is not seen in the sex cells, it might still occur during embryogenesis—the process by which a zygote becomes an embryo. These changes, which are referred to as germline mutations in both situations, are heritable by an individual's progeny.

6.2.2. Genetic Alterations, Mutations, and Polymorphisms

6.2.2.1. Mutations

The term "mutations" is probably associated with something unfavorable in your mind. This makes sense because mutations, which are genetic changes associated with a higher risk of disease development, are found in fewer than 1% of the population.

Take the *BRCA1* gene, which you have undoubtedly heard of. This gene regulates cell division appropriately to keep tumors from developing. An increased chance of developing a tumor is caused by an uncontrolled division of cells caused by a mutation in this gene. Stated differently, individuals with a *BRCA1* mutation are 46–87% more likely to develop breast cancer in their lifetime.

6.2.2.2. Polymorphisms

Genetic changes known as polymorphisms affect more than 1% of the population. The majority of polymorphisms are single nucleotide polymorphisms, or SNPs, which are genetic variations that only impact the substitution of one nucleotide for another. Currently, the genome contains millions of SNPs that are known to exist. As a matter of fact, the genome is thought to contain one SNP for every 100 to 1000 bases (A, T, G, and C).

Traits phenotypic, i.e., genetic polymorphisms determine outward characteristics that set us apart from one another, like eye color and height. The majority of SNPs (98% of DNA) are located in non-coding regions and have no direct impact on health. A person's susceptibility to a specific multifactorial disease, which is influenced by both genetic and environmental factors, can be affected by additional SNPs found in coding regions (2 percent of DNA).

6.2.2.3. Genetic Variation, the Key to Evolution

Therefore, what makes each of us unique are these genetic variations found in our genes. Since mutations are the source of all genetic variation, there couldn't be evolution without genetic variation either i.e., stable and inherited genetic alterations that occur (in successive generations). Due to their random occurrence, mutations boost genetic diversity but serve no adaptive purpose. Natural selection has regulated each species' unique rate of mutation to enable it to adapt in a balanced manner to the dualities of stability and change present in every environment.

Did you get it?

1. **Why do genetic alterations occur?**
2. **What do you understand by polymorphisms?**

6.3. EXISTENCE OF GENETIC VARIATION AMONG HUMANS

Compared to the great majority of species on Earth, the majority of which exist for much longer than humans, Homo sapiens is a relatively young species and has not had as much time to accumulate genetic variation. However, our species exhibits a great deal of genetic variation. With the exception of identical twins, no two humans have ever been or will ever be genetically identical. The human genome is made up of roughly 3 × 109 base pairs of DNA. The genetic variation, or biochemical individuality, between any two humans is only a few percent. This implies that any two people will differ by roughly one base pair out of every 1000. The difference between any two (diploid) individuals is approximately 6×10^6 base pairs, which is a major factor in the development of automated methods to examine genetic variation.

In the human genome, single base-pair differences are the most prevalent type of polymorphisms or genetic variations. These variations are referred to as single-nucleotide polymorphisms, or SNPs, by scientists. Comparing two distinct haploid genomes reveals that SNPs are present, on average, every 1000 bases. Although they happen far less frequently, other kinds of polymorphisms such as variations in copy number, insertions, deletions, duplications, and rearrangements also happen.

Even though people differ genetically from one another, most human genetic information is shared by all. These commonalities aid in defining our species. Furthermore, there are no distinct, discontinuous borders between the various human population groups; rather, genetic variation is distributed globally in a rather continuous manner. As a matter of fact, studies have consistently shown that, while only 15% of human genetic variation is found between populations, 85% of it is found within human populations. In other words, evidence suggests that Homo sapiens is a single, constantly changing species that interbreeds. Even traditional ideas about human racial groups have been called into question by biologists and physical anthropologists as a result of ongoing research into human genetic variation. In fact, the genetic variation between these conventional taxonomies is less than that which taxonomists use to classify subspecies—that is, the taxonomic category for other species that is equivalent to the classification of race in humans (Figure 6.10). Some biologists have seriously questioned the validity of race as a biological construct in light of this finding.

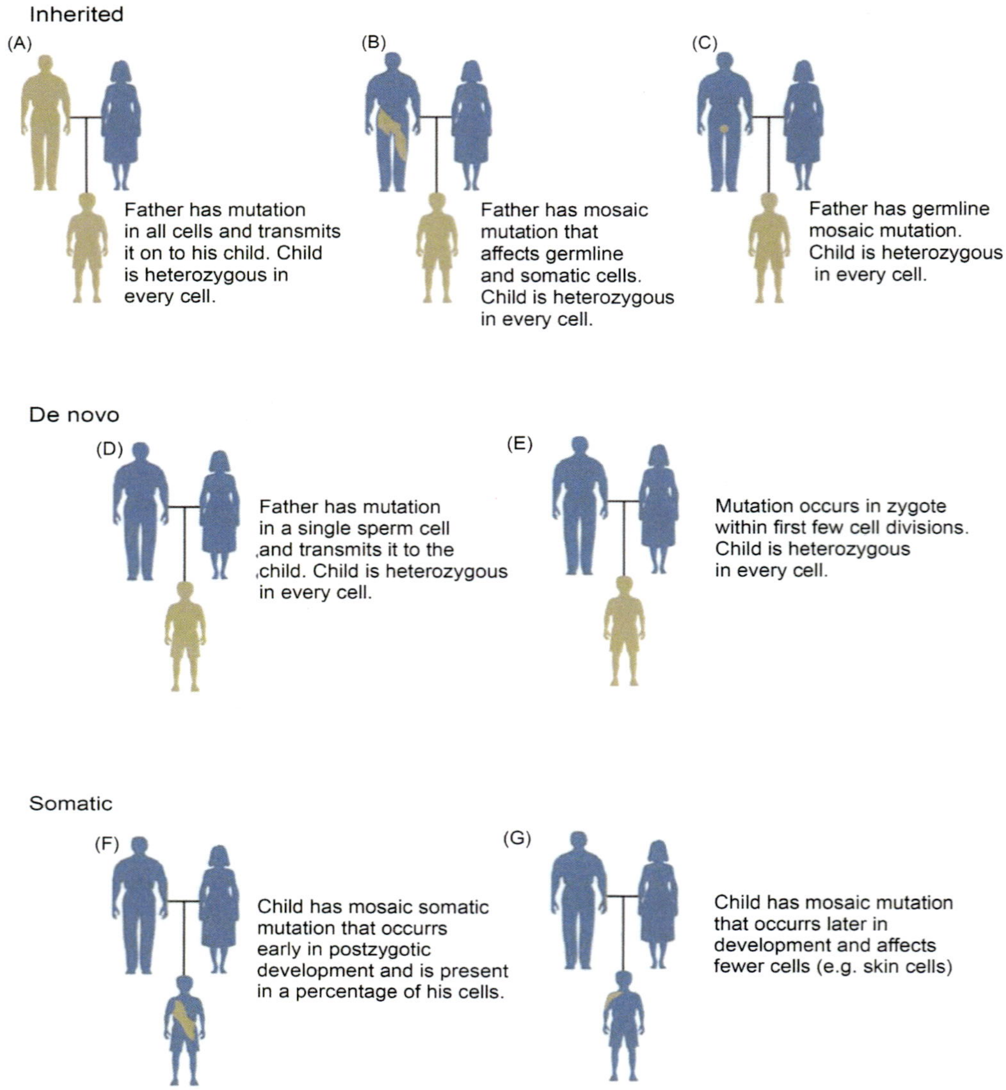

Figure 6.10. *Most variation occurs within populations.*

Source: Freed, D., Stevens, E. L., & Pevsner, J., (2014). Somatic mosaicism in the human genome. Genes, 5(4), 1064–1094. https://doi.org/10.3390/genes5041064.

The analysis of genetic variation in humans also reveals that a large portion of human DNA is shared by other members of the natural world, indicating that all life is related through descent with modification from common ancestors. Particularly in development-related genes, the highly conserved nature of many genetic regions across a great evolutionary distance is evident. For instance, mutations in the patched homolog in humans cause structural abnormalities in the developing human embryo, whereas mutations in the patched gene cause developmental abnormalities in Drosophila.

The fact that evolution is conservation has allowed geneticists to identify genetic variants linked to certain cancers. For instance, one type of colon cancer is linked to mutations in the genes that fix DNA mismatches that occur during DNA replication. These mismatched repair genes, known as Mutl and Muts, have been conserved throughout evolutionary history, all the way back to the bacterium *Escherichia coli*. Utilizing the sequences that are known from its chromosome, geneticists surmised that this type of colon cancer was linked to a mismatch repair failure *E. coli* genes to search for homologous sequences in the human genome. The culmination of this research was the discovery of a gene linked to a higher risk of colon cancer.

Did you know?

In 1971, Berg utilized restriction enzymes to create the first recombinant DNA molecule and the first recombinant DNA plasmid. In 1972, Cohen and Boyer created the first recombinant DNA organism by inserting recombinant DNA plasmids into E. coli, now known as bacterial transformation, and paved the way for molecular cloning.

6.4. SIGNIFICANCE OF HUMAN GENETIC VARIATION

Learning Objectives

- Discuss about health and disease susceptibility
- Explain the personalized medicine
- Describe the human evolution and adaptation
- Focus on forensics and genetic identification
- Learn about diversity in traits and abilities

In terms of biology, almost all human genetic variation is negligible; that is, it has no bearing on adaptation. A neutral mutation, for instance, modifies the resulting protein's amino acid sequence without appreciably changing its function. Other variations don't even alter the amino acid sequence, like a silent mutation. Moreover, only a tiny portion of the DNA sequences in the human genome are regulatory sequences, which control the timing, degree, and tissue specificity of gene expression, or coding sequences, which are sequences that are eventually translated into proteins. There is no effect from variations that exist elsewhere in the DNA, which makes up the great majority of the DNA with no known function.

On the other hand, some genetic variation may be advantageous in dynamic environments. The sickle hemoglobin mutation, which in its heterozygous state confers a selective advantage in regions where malaria is endemic, is a well-known example from high school biology curricula.

Examples from more recent times include changes in the CCR5 gene, which seem to offer a defense against AIDS. A protein on the surface of human immune cells is encoded by the CCR5 gene. HIV, the virus that causes AIDS, attaches itself to this protein as well as another protein on the surface of immune cells to infect them. HIV infection can be reduced by mutations in the CCR5 gene that change the protein's structure or the degree to which it is expressed. Early studies on a particular genetic variant suggest that it might have reached high frequency in Northern Europe approximately 700 years ago, coinciding with the European bubonic plague epidemic. This discovery has prompted some scientists to speculate that the CCR5 mutation might have offered a defense against *Yersinia pestis* infection, which causes plague. The fact that HIV and *Y. pestis* infect macrophages lends credence to the genetic variant's selective advantage.

The reports of sickle cell AIDS and plague serve as a reminder that the environment in which genes are expressed determines the biological significance of genetic variation. It also serves as a reminder that evolution and differential selection cannot occur in the absence of genetic variation within a species.

Of course, there is a genetic component to disease; classic single-gene disorders like Duchenne muscular dystrophy, sickle cell disease, and cystic fibrosis serve as reminders of this. Research is also beginning to identify genetic variations linked to common diseases that are among the leading causes of illness and mortality in developed nations: heart disease, cancer, diabetes, and psychiatric disorders like bipolar disorder and schizophrenia (manic-depressive illness). The more prevalent diseases are caused by the interaction of several genes and environmental factors, while disorders like Huntington's disease and cystic fibrosis are caused by mutations in a single gene and are visible in almost every environment. Therefore, these illnesses are referred to as multifactorial and polygenic. In actuality, most human characteristics—diseases included—are multifactorial.

There are important genetic differences between more common multifactorial diseases and relatively rare single-gene disorders. The single-gene disorders caused by these genetic variations are usually quite recent, and they frequently have a significant negative effect that seriously disturbs homeostasis. These illnesses also usually take a toll early in life, frequently before childhood ends. On the other hand, common multifactorial diseases are typically caused by older genetic variations that affect homeostasis more gradually and to a lesser extent. Additionally, they typically start in adulthood. Given the last two features, even in the presence of a genetic predisposition, people can alter their behavior in ways that decrease the chance of developing a common disease. This makes the ability to identify genetic variations that predispose or increase the risk of common diseases particularly valuable.

Human genetic variation has profound significance across various fields, from medicine to evolution. The importance of genetic diversity in human populations:

6.4.1. Health and Disease Susceptibility

Genetic diversity is a major factor in defining a person's vulnerability to various illnesses. Gene variations can affect a person's risk of developing diseases like cancer, diabetes, cardiovascular disease, and infectious infections.

- **Single Gene Disorders**: Mutations in single genes can lead to hereditary diseases like cystic fibrosis, sickle cell anemia, and Huntington's disease.
- **Polygenic Diseases**: Most common diseases, such as type 2 diabetes, hypertension, and heart disease, are influenced by variations in multiple genes. Individuals with certain genetic variants may be more prone to these diseases.
- **Infectious Diseases**: Genetic variation can also affect how individuals respond to infections. For example, certain mutations can

offer resistance to diseases like malaria (e.g., the sickle cell trait) or HIV (e.g., the CCR5-Δ32 mutation).

6.4.2. Personalized Medicine

For personalized medicine to advance, wherein treatments are customized based on a patient's genetic profile, a basic understanding of genetic variation is required. Genetic differences can affect drug metabolism, efficacy, and risk of adverse reactions, resulting in a variety of responses to drugs among individuals.

- **Pharmacogenomics**: This field studies how genetic variations affect drug responses. For instance, some people metabolize drugs faster or slower, meaning that standard doses may not be appropriate. Genetic testing can help doctors choose the right medication and dosage for each patient.
- **Cancer Treatment**: Many cancers are driven by specific genetic mutations. By identifying these mutations, targeted therapies can be developed that are more effective and less toxic than traditional treatments.

6.4.3. Human Evolution and Adaptation

Evolution's starting point is genetic variation. Through the process of natural selection, it enables populations to adjust to changing environments. Over time, evolutionary changes can be attributed to variations in genes that affect survival and reproduction in particular environments.

- **Skin Color**: Variations in genes like MC1R and SLC24A5 have contributed to differences in skin pigmentation. These changes are thought to have evolved in response to varying levels of ultraviolet radiation in different parts of the world.
- **Lactose Tolerance**: The ability to digest lactose (a sugar in milk) in adulthood is due to genetic variation in the **LCT** gene. This trait is more common in populations with a long history of dairy consumption, demonstrating human adaptation to agricultural practices.
- **Altitude Adaptation**: Populations living at high altitudes, such as Tibetans and Andeans, have genetic adaptations that allow them to thrive in low-oxygen environments. These adaptations involve variations in genes related to oxygen transport and metabolism.

6.4.4. Forensics and Genetic Identification

DNA analysis allows for individual identification, which makes genetic variation essential to forensic science. To identify suspects, victims, and relatives, forensic scientists can use DNA testing because each person (apart from identical twins) has a unique genetic composition.

- **DNA Fingerprinting**: This technique relies on analyzing specific regions of the genome that vary greatly between individuals. These regions, such as short tandem repeats (STRs), are used in criminal investigations, paternity tests, and identifying human remains.
- **Population Studies**: Genetic variation helps trace human ancestry and migration patterns. By studying genetic markers across populations, scientists can track how humans have spread across the globe over time.

6.4.5. Diversity in Traits and Abilities

Genetic variation accounts for the wide range of human physical traits and abilities, such as height, eye color, hair type, athletic performance, and cognitive abilities.

- **Physical Appearance**: Traits like hair texture, eye color, and facial features are influenced by multiple genes, which vary across individuals and populations. This contributes to the rich diversity of human appearances.
- **Cognitive and Behavioral Traits**: While heavily influenced by the environment, genetic variations also play a role in intelligence, personality, and susceptibility to mental health conditions like depression and schizophrenia.

6.4.6. Conservation of Genetic Diversity

Maintaining genetic variation in populations is essential for long-term survival from an evolutionary standpoint. Because some members of a genetically diverse population may possess variations that enable them to survive and procreate in harsh environments, this population is more resistant to diseases, environmental changes, and other threats.

- **Bottleneck Effect**: When populations experience drastic reductions in size (bottlenecks), genetic variation is lost, which can reduce the population's ability to adapt to new environments or resist diseases.
- **Founder Effect**: This occurs when a new population is established by a small number of individuals, leading to reduced genetic variation. Such populations may be more vulnerable to genetic diseases and environmental changes.

6.4.7. Cultural and Social Impacts

Cultural and social perspectives on race and ethnicity have been impacted by the study of human genetic variation. According to genetic research, human populations are not biologically distinct as suggested by racial categories. Rather, there is little genetic variation between populations, and the majority of genetic variation occurs within populations. The fight against racism and discrimination will be significantly impacted by this scientific understanding.

Diversity, health, and the evolution of human populations all depend on genetic variation. It affects physical characteristics, adaptation to environmental changes, medication responses, and disease risk. Understanding genetic variation can help advance medical treatments, shed light on human evolution, and address issues in society pertaining to diversity and identity as long as human genetics research is conducted.

Did you get it?

1. Discuss the conservation of genetic diversity
2. Explain the cultural and social impacts

6.5. HUMAN GENETIC VARIATION AND MEDICINE

Learning Objectives

- Discuss about causes of variation
- How to measures of variation

Understanding human genetic variation is useful for both understanding and promoting health as well as for understanding and combating disease, as was previously mentioned. It is unlikely that we can overstate how significant this advantage is. First, almost all human diseases have a genetic component, as Figure 6.11 illustrates. This component is very large in some diseases, like Tay-Sachs disease, Huntington's disease, and cystic fibrosis. There is less of a genetic component in other diseases like diabetes, cancer, and heart disease. Since we only inherit a predisposition to develop a disease rather than the certainty of developing it, we actually do not typically think of these conditions as "genetic diseases."

There is very little genetic component to still other diseases. But the important thing to remember is that it exists. There is a true, albeit tiny, genetic component to even infectious diseases, which we have historically classified as entirely distinct from genetic disorders. For instance, AIDS itself is impacted by an individual's genotype, as demonstrated by the CCR5 example mentioned previously. In fact, because they carry a variant of the CCR5 gene, some people seem to have a genetic resistance to HIV infection.

Second, since everyone has some genetic risk, we can all theoretically gain from the advancements scientists are making in our knowledge of these risks and how to mitigate them. Each of us is thought to have between 5 and 50 mutations that increase our chance of developing a disease or disability. Because we do not live long enough for the consequences of our mutations to manifest or because we are not exposed to the necessary environmental triggers, some of us may not suffer negative effects from the mutations we carry. But the truth is that each of us carries the possibility of unfavorable outcomes arising from our genetic makeup.

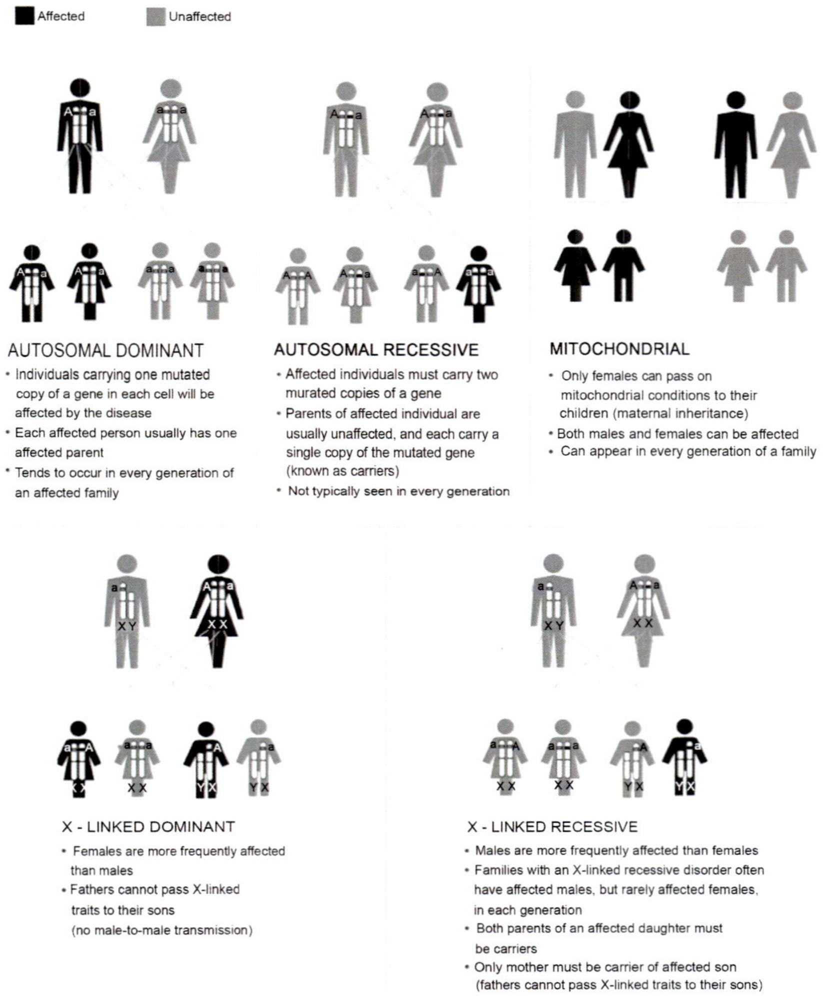

Figure 6.11. *Virtually all human diseases, except perhaps trauma, have a genetic component.*

Source: Genetic Alliance; The New York-Mid-Atlantic Consortium for Genetic and Newborn Screening Services. Understanding Genetics: A New York, Mid-Atlantic Guide for Patients and Health Professionals. Washington (DC): Genetic Alliance; Chapter 1, Genetics 101. Available from: https://www.ncbi.nlm.nih.gov/books/NBK115568/.

As Figure 6.12 illustrates, modern genetic analysis of a human disease starts with mapping and cloning the associated gene or genes. This is one way that modern genetics is helping us address the challenge of human disease. The genes linked to retinoblastoma, cystic fibrosis, and Duchenne muscular dystrophy were among the first disease genes to be mapped and cloned. Recently, researchers have revealed that the genes for diabetes, Parkinson's disease, and breast cancer have been cloned.

The mapping and cloning of a disease-related gene facilitates the creation of numerous novel health care approaches, as Figure 6.12 further demonstrates. Genetic tests that aim to identify individuals who are more susceptible to the disease and identify genotypic variations that may affect the efficacy of treatment are at one extreme of the spectrum. On the other extreme are novel drug and gene therapies that selectively target the biochemical pathways underlying the symptoms of the disease or even substitute, modify, or augment defective genes with functional ones. In fact, as Figure 6.12 indicates, the era of molecular medicine is upon us.

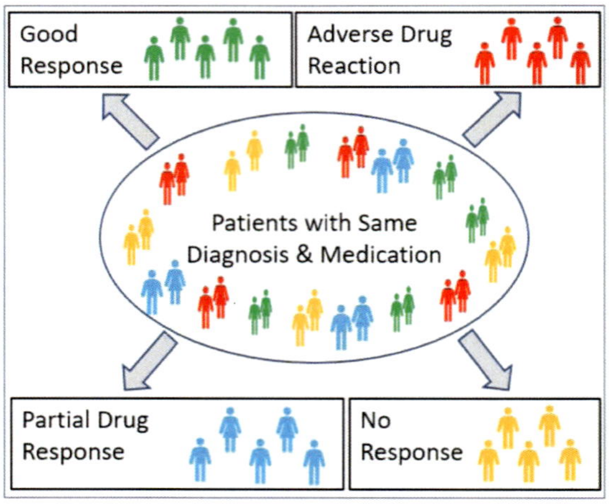

Figure 6.12. *Mapping and cloning a gene can lead to strategies that reduce the risk of disease (preventive medicine); guidelines for prescribing drugs based on a person's genotype (pharmacogenomics); procedures that alter the affected gene (gene therapy).*

Source: Gill, P. S., Yu, F. B., Porter-Gill, P. A., Boyanton, B. L., Allen, J. C., Farrar, J. E., Veerapandiyan, A., Prodhan, P., Bielamowicz, K. J., Sellars, E., et al., (2021). Implementing pharmacogenomics testing: Single center experience at Arkansas Children's Hospital. Journal of Personalized Medicine, 11(5), 394. https://doi.org/10.3390/jpm11050394.

Genetic testing is not a brand-new approach to medical care. In many states, newborn screening for conditions like PKU has been in place for 30 years. However, the incredible strides being made in the mapping and cloning of human disease genes offer hope for the future creation of additional genetic tests. The availability of these tests can alter the way that the general public views a given illness and the type of care that members of affected families may seek and receive. For instance, the discovery of the *BRCA1* and *BRCA2* genes and the proof that specific variations of these genes are linked to a higher risk of ovarian and breast cancer have made it possible to establish guidelines and testing procedures for people with a family history of these illnesses. The first gene to be isolated, *BRCA1*, is found on the long arm of chromosome 17. Variants of this gene cause roughly half of all inherited breast cancers or 5% of all cases of breast cancer. Approximately 30 to 40% of all inherited breast cancers seem to be caused by variants of *BRCA2*, which are found on the long arm of chromosome 13.

According to scientific estimates, 1 of the hundreds of thousands of significant mutations in the *BRCA1* gene that have already been identified affects women in the United States. Knowing that she carries one of the *BRCA1* or *BRCA2* variants linked to an increased risk can be crucial information for a woman with a family history of breast cancer. In the event that she carries one of these variants, she and her doctor may decide to modify her care in a number of ways, including having a prophylactic mastectomy, having more frequent physical exams, and starting mammography earlier in life. In the future, medications that lower the risk of breast cancer development might also be accessible.

The way medications are developed and prescribed is also evolving due to the capacity to test for the presence of specific gene variants in individuals. Pharmacogenomics, a fast-expanding field, focuses on important genetic variations that make medications function better in some individuals and less effectively—or with potentially harmful side effects—in others. For instance, the presence or absence of one of three genetic variations in the ApoE (Apolipoprotein E) gene can affect how patients react to medication treatment, according to research on the disease. Likewise, genetic variations in the beta-2-adrenergic receptor have been recently connected to some of the variability in kids' reactions to therapeutic dosages of albuterol, a medication used to treat asthma. Given that the most commonly used agents for treating asthma are beta-2-adrenergic receptor agonists, such as albuterol, these findings may have significant ramifications for our comprehension of the genetic variables influencing an individual's response to asthma treatment.

Experts believe that in the future, doctors will more often use genetic testing to match medications to the unique body chemistry of each patient, allowing them to prescribe the safest and most effective medications at the right dosages. Pharmaceutical companies will probably start working on developing new, highly specific drugs and reviving older ones whose effects previously seemed too unpredictable to be of clinical value after determining the genotypes that determine individual responses to specific drugs.

Understanding the molecular makeup of genes linked to disease is also altering how scientists approach creating novel medications. One notable instance came after the 1989 identification of the gene linked to cystic fibrosis (CF). To comprehend the biochemical effects of the variant forms of the gene and to create new treatment plans based on that understanding, researchers started examining the functions of the normal and defective proteins involved. The normal protein, known as the cystic fibrosis transmembrane conductance regulator, or CFTR, is embedded in the membranes of various body cell types. It functions as a channel in these cell membranes, facilitating the removal of chloride ions. Depending on the specific mutation a patient has, the CFTR protein in the cell membrane may be absent, reduced, or both in CF patients. It may also be present but not functioning correctly. Certain mutations cause the CFTR protein to not synthesize at all, leaving the cells devoid of CFTR molecules.

While every CF mutation affects chloride transport, the effects on individual patients differ depending on the specific mutation. Patients with mutations that result in either a marked reduction in the amount of CFTR protein or in its absence may, for instance, have a more severe form of the disease than patients whose CFTR protein is present but functions differently. Diverse treatment approaches are also suggested by the various mutations. For instance, the most prevalent mutation associated with CF, known as delta F508, results in the synthesis of protein molecules, known as delta F508 CFTR, which are improperly processed and break down before they can reach the cell membrane. Based on this finding, patients with delta F508 CFTR may benefit greatly from medication treatments that improve the transport of the defective delta F508 protein to the cell membrane or stop its degradation.

Ultimately, the discovery, replication, and sequencing of a gene linked to a disease may pave the way for the creation of therapeutic approaches based on the instructions contained within the gene. These tactics, together known as gene therapy, usually entail introducing a copy of the healthy version of a gene linked to a disease into the cells of a patient. The most well-known instances of this kind of gene therapy involve transplanting a patient's cells back into their body to replace a missing gene, using a vector to introduce the healthy variant of the disease-related gene. The adenosine deaminase (ADA) gene's normal allele was first inserted into the body of a young girl who had been born with ADA deficiency using this technique in the early 1990s. An aberrant form of the ADA gene causes this illness by impairing the production of adenosine deaminase, a protein necessary for T-lymphocytes to function properly.

While scientists are still working to improve this broad strategy for gene therapy, they are also creating new strategies. For instance, researchers are hopeful that chimeraplasty, a relatively novel tactic, will one day be able to effectively fix genetic flaws involving just one base change. In order to replace an incorrect base with the correct one, chimeraplasty employs specifically designed molecules that base pair with the patient's DNA and activate the cell's natural DNA repair processes. Chimeraplasty is still in its infancy as of right now, and the first clinical trials will soon begin.

Introducing new genetic information into a cell to give it new or different functions is another method of gene therapy. For instance, recent research has shown that genetic material can be introduced into cancer cells under strictly regulated experimental settings, changing their metabolism to the point where the cells kill themselves in response to an environmental trigger that would ordinarily be harmless. Similar tests are being conducted by researchers to examine the viability of introducing genetic modifications into cells that will render them resistant to HIV infection. Even though this research is limited to nonhuman primates for the time being, it may eventually help HIV-positive patients.

The Human Genome Project (HGP) has greatly accelerated the rate of human gene discovery as well as the creation of novel health care approaches predicated on an understanding of a gene's structure and function. The expense of locating human genes is also being decreased by the new insights and technological advancements arising from HGP-related research. For example, it took more than eight years and $50 million to find the gene linked to cystic fibrosis; this search was completed in 1989, prior to the creation of the HGP. On the other hand, the discovery of a gene linked to a Mendelian illness can now be completed for about $100,000 and in less than a year.

In recent years, there has also been a gradual shift in the focus of genetic variation research on humans from primarily studying genes linked to single-gene disorders—which are relatively uncommon in the human population—to a growing number of genes linked to multifactorial diseases. We can anticipate that many more people will be impacted by this work because these diseases are not uncommon. Comprehending the genetic and environmental origins of these complex illnesses will also prompt more investigation and the creation of novel treatments, which will probably have a significant impact on medical practice in the coming century.

6.5.1. Causes of Variation

Individual differences are caused by independent assortment, different mutational events, and gene exchange (crossing over and recombination) during meiosis, the process of reproduction. There are genetic variations among populations for at least three reasons. If a particular allele gives a competitive advantage, natural selection may confer an adaptive advantage on individuals in that particular environment. Only those geographical areas where they provide an advantage are likely to have alleles under selection. The result of haphazard alterations in the gene pool under circumstances where the majority of mutations are neutral—that is, they do not seem to have any discernible beneficial or negative selective effects on the organism—is genetic drift, which is a second significant mechanism. In conclusion, there exist statistical disparities between small migrant groups and the larger populations from which they originated. This phenomenon is known as the "founder effect."

Remember

When these migrants migrate to new areas, their offspring populations usually diverge from the original populations in that they have a lower genetic diversity and a dominant subset of different genes.

The primary cause in humans is genetic drift. Neutral differences between populations may have been significantly influenced by past small population sizes and serial founder effects, which raise the possibility of genetic drift. The majority of mutations have a high degree of neutrality, which is the second primary cause of genetic variation. A notable proportion of genes seem to have been subjected to recent natural selection, with selective pressures occasionally being region-specific.

6.5.2. Measures of Variation

Human genetic variation can take many different forms, ranging from significant modifications to single nucleotide shifts in the karyotype. Out of 160 live human births, one has a chromosomal abnormality. The majority of aneuploidy cases, aside from sex chromosome disorders, result in miscarriages, which are the death of the developing fetus; the extra autosomal chromosomes that are most frequently found in live births are 21, 18, and 13.

The average percentage of nucleotides that split two people apart is called nucleotide diversity. The estimated nucleotide diversity of humans as of 2004 ranged from 0.1 to 0.4 percent of base pairs. The reference human genome and an individual's genome differ at 4:1 million to 5:0 million sites, affecting 20 million bases of sequence (the latter representing 0:6% of all base pairs). This was discovered in 2015 by the 1000 Genomes Project, which sequenced 1000 individuals from 26 human populations.

SNPs and indels account for a smaller number of base pairs than structural variations, which make up nearly all (>99.9%) of these sites, which are small variations in the genetic sequence that are either single nucleotide polymorphisms or brief insertions or deletions (indels). 324 million variants were identified in sequenced human genomes as of 2017, according to the Single Nucleotide Polymorphism Database (dbSNP), which catalogs SNP and other variants.

6.5.2.1. Single Nucleotide Polymorphisms

A single nucleotide variation that affects at least 1% of a population's members is known as a single nucleotide polymorphism, or SNP. 84.77 million SNPs were found in the 2,504 people that the 1000 Genomes Project studied. 90% of all sequence variants are thought to be SNPs, making them the most prevalent type of sequence variation. This was estimated in 1998. Single base swaps, deletions, and insertions are examples of additional sequence modifications. Generally occurring every 100 to 300 bases, single nucleotides (SNPs) are the primary cause of heterogeneity.

A functional SNP, also known as a non-synonymous SNP, is one that modifies messenger RNA or gene splicing and, as a result, alters phenotypic differences within the species. Human SNPs with functional frequencies range from 3 to 5 percent (International HapMap Project). Because of their large number and steady inheritance over many generations, neutral or synonymous SNPs are still valuable genetic markers in genome-wide association studies. One that is found inside a gene is known as a coding SNP. 103 genes have premature stop codons as a result of 105 Human Reference SNPs. This is equivalent to 0.5 percent of SNPs that code for coding. They arise from the genome's segmental duplication. Although all of these SNP alleles are common and do not undergo purification through negative selection, these SNPs cause a loss of protein.

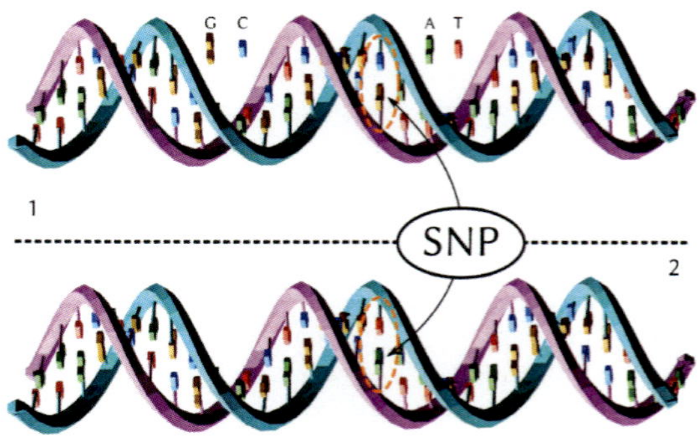

Figure 6.13. *DNA molecule 1 differs from DNA molecule 2 at a single base-pair location (a C/T polymorphism).*

Source: https://upload.wikimedia.org/wikipedia/commons/thumb/2/2e/Dna-SNP.svg/780px-Dna-SNP.svg.png.

6.5.2.2. Structural Variation

Variations in an organism's chromosome structure are known as structural variation. The diversity of single nucleotides is far less common in human genetic variation than structural variations, which include copy-number variation and deletions, inversions, insertions, and duplications. This was determined in 2007 through examination of the diploid complete sequences of the genomes of James D. Watson and Craig Venter. This supplemented the two haploid sequences, release by Celera Genomics and the Human Genome Project, respectively, that were assemblages of sequences from multiple individuals. A normal human has 2,100–2,500 structural variations, including about 1000 large deletions, 160 copy-number variants, 915 Alu insertions, 128 L1 insertions, 51 SVA insertions, 4 NUMTs, and 10 inversions, according to the 1000 Genomes Project.

- **Copy Number Variation:** A chromosome's copy-number variation (CNV) is a change in the genome brought about by the duplication or deletion of substantial sections of DNA. The copy number of unrelated human genomes differs in an estimated 0.4 percent of cases. The estimated human-to-human genetic variation is at least 0.5 percent (99.5% similarity) when copy number variation is taken into account. Variations in copy numbers can occur during development as well as be inherited.
- **Epigenetics:** Variations in the chemical labels that bind to DNA and influence **gene** reading are known as epigenetic variations. The labels, also known as epigenetic markings, function as switches to regulate the reading of genes. Certain alleles allow the epigenetic state of the DNA and the phenotypic that goes along with it to be passed down through generations of people.
- **Genetic Variability:** The tendency of individual genotypes within a population to vary (become different) from one another is measured as genetic variability. Genetic diversity, or the amount of variation present in a given population, is not the same as variability. The degree to which a trait tends to change in response to hereditary and environmental factors is known as its variability.
- **Clines:** In the realm of biology, a cline represents a seamless progression of species, populations, variations, or organism forms, displaying progressive phenotypic and/or genetic alterations, spread across a geographical expanse. This often occurs due to the diverse environmental conditions present. In the field of human genetic research, a gene cline can be precisely outlined and measured quantitatively.
- **Haplogroups:** A haplogroup, as used in molecular evolution research, is a collection of related haplotypes that have a single nucleotide polymorphism (SNP) mutation in common with a common ancestor. Haplogroup research offers insights into the ancestry of ancestors that go back thousands of years.

Since they can both be used to define genetic populations, Y-chromosome (Y-DNA) and mitochondrial DNA (mtDNA) haplogroups

Keyword

A gene is a segment of DNA that is the basic unit of heredity, passed from parents to offspring.

are the most frequently studied human haplogroups. While mtDNA is passed down the matrilineal line from mother to both daughter and son, Y-DNA is only passed along the patrilineal line, from father to son. Every generation may see a random mutation that modifies the mtDNA and Y-DNA.

Variable Number Tandem Repeats: A tandem repeat's length variation is called a variable number tandem repeat (VNTR). A tandem repeat is a brief nucleotide sequence that is repeated next to one another. Numerous chromosomes contain tandem repeats, which differ in length amongst people. Since each variation functions as an inherited allele, they can be used to identify a person or their family. Their analysis has applications in DNA fingerprinting, forensics, and biology and genetics research. Microsatellites are short tandem repeats with five base pairs or less, and minisatellites are longer tandem repeats.

Did you get it?

1. **Focus on single nucleotide polymorphisms**
2. **Describe the structural variation**
3. **What do you understand by haplogroups?**

6.6. GENETICS, ETHICS, AND SOCIETY

Learning Objectives

- Discuss the genetic privacy and confidentiality
- Focus on genetic discrimination
- Explain the genetic testing and reproductive choices

As previously mentioned, the speed at which genetic factors in disease are being discovered has improved our ability to predict the risk of disease in asymptomatic individuals. But what are the implications of applying our expanding understanding of human genetic variation to improve personal and public health? However, there is still a great deal to learn about the advantages and drawbacks of developing a molecular understanding of human genetic variation. The HGP's designers recognized that while this knowledge might significantly improve human health, it would also bring up a number of difficult moral, legal, and social issues. In order to foresee and handle the ethical, legal, and social issues that result from human genetic research, they founded the Ethical, Legal, and Social Implications (ELSI) program in 1990. Perhaps more than any other program, this one has drawn attention from the general public and educators to the growing significance of educating the public so that they can participate in and comprehend the ongoing public discourse surrounding genetic advancements.

The study of good and evil, right and wrong, is known as ethics. It is related to the behaviors and personalities of people as well as families, communities, organizations, and societies. For the past two and a half millennia, Western philosophy has produced a plethora of effective techniques as well as a trustworthy lexicon of technical terms and concepts for the study and discussion of ethical life. In general, we define "right" and "good" as behaviors and attributes that advance the interests of people as individuals, families, communities, institutions, and as a whole. In this context, a participant's share or involvement in a situation is referred to as their "interest." "Wrong" or "bad" refers to behaviors and characteristics that harm one's interests.

The issues surrounding ethical considerations are numerous, intricate, and complicated. Concerning what constitutes good and bad behavior or actions of an individual or group, there are frequently opposing, well-reasoned opinions. All of these responses typically make reference to moral principles. Something that is significant or valuable in a particular circumstance is called a value. The diverse ways that the participants assign values to objects, people, and situations are among the most fascinating things to observe during any ethics discussion. In an ethical discussion, students could bring up values like autonomy, freedom, privacy, the sanctity of life, religion, defending the rights of others, advancing their own welfare, justice, fairness, relationships, science, and technological advancement.

Recognizing the intricate and diverse aspects of ethical conversations does not imply that "anything goes." Experts concur on the following aspects of ethics in general. To begin with, ethics is a logical inquiry process. It entails asking thoughtful questions and looking for thoughtful responses to those questions. For instance, we can inquire about the appropriateness of specific gene therapy applications as well as an individual's right to privacy with regard to their genetic information. Arguments are well-reasoned responses to these kinds of queries. Thus, an ethical investigation that is successful leads to ethical analysis and reasoning.

Keyword

Gene Therapy – It is a technique that uses a gene(s) to treat, prevent or cure a disease or medical disorder.

Second, a strong information base and careful interpretation of that data are necessary for ethics. To assess the recent move by the Icelandic government to establish a database containing a wealth of genetic and medical data about its citizens, for instance, one needs to have a strong background in biology. Speaking about the morality of germ-line gene therapy or genetic screening requires a scientific background as well. Ethics is not just a theoretical field; it also critically addresses practical issues.

Third, debating ethical matters frequently reveals a wide range of responses to queries concerning what is good and bad as well as what is right and wrong. This is particularly true in a society like ours where there is a wide range of viewpoints and values. For instance, think about the issue of testing teenagers for late-onset genetic disorders. Genetic testing facilities frequently deny asymptomatic patients under the age of 18 access to genetic testing for Huntington's disease (HD). The reasoning behind this is that there is currently no treatment available and the condition manifests itself later in life. Therefore, a specific diagnosis based on genetic testing does not immediately improve a minor's physical health. Concerns have also been raised regarding the psychological fallout from knowing that one will eventually develop a crippling, potentially fatal illness. Teens are free to postpone deciding what they want to know and when until they are adults. Some respond by saying that many young children and adolescents do have enough autonomy to give their consent and make decisions, and they might even want to know what lies ahead for them. Some contend that parents ought to be granted the privilege.

This module aims to teach students how a scientific understanding of the world can support individuals and society in making informed decisions about genetics and health-related issues. In Activity 5, "Making Decisions in the Face of Uncertainty," students are given the example of a woman who worries that she might have a mutated gene that makes her more likely to develop ovarian and breast cancer. Students also think about the various decisions that the woman must make. Thus, prudential decision-making—the capacity to shun needless risk when it is unclear whether an event will truly occur—is the main focus of Activity 5. By completing the exercise, students gain an understanding of why questions about genetics and health frequently involve uncertainty because our understanding of genetics is still developing and incomplete. Students also realize how difficult it is to make decisions when faced with an uncertain future. Simply put, when the issues are weighed, different outcomes are possible, depending on one's estimate of the incidence of the occurrence and how much burden one attaches to the risk. Students should ask themselves, "How bad is the outcome and how likely is it to occur?"

It is obvious that both science and ethics are crucial in assisting people in making decisions regarding their own and the public's health. Science offers data that can aid in our understanding of and ability to treat human dysfunction, disease, and deformity. Furthermore, ethics offers a framework for defining and making sense of values and the decisions that follow from them. However, there are complications in the relationships that exist between scientific knowledge and human decisions as well as between decisions and behaviors. To put it another way, people have the freedom to make decisions that contradict solid information and do not necessitate taking action.

The ethical and societal issues related to human genetic variation:

6.6.1. Genetic Privacy and Confidentiality

Human genetic variation research often involves the collection of sensitive genetic data from individuals and populations. Protecting the privacy and confidentiality of this data is a major ethical concern.

- **Data Sharing**: Advances in genomics often require large-scale data sharing across institutions and countries. This raises concerns about who controls access to genetic information and how to ensure it is not misused.
- **Potential Misuse**: Genetic information could be used in ways that harm individuals, such as revealing predispositions to diseases to insurance companies or employers, potentially leading to discrimination.

6.6.1.1. Ethical Considerations

- Ensuring informed consent for genetic testing, particularly when participants may not fully understand the implications of sharing their genetic information.
- Establishing clear guidelines on how genetic data can be used, who has access to it, and how it is stored to protect against breaches of confidentiality.

6.6.2. Genetic Discrimination

People who are treated unfairly because of their genetic makeup are engaging in genetic discrimination. Concern over the possibility of discrimination in the workplace, insurance market, and healthcare is growing as genetic information becomes more widely available.

- **Health Insurance**: Individuals with a genetic predisposition to certain diseases may be denied coverage or charged higher premiums.
- **Employment**: Employers might use genetic information to avoid hiring individuals who are likely to develop costly health conditions, even if they are currently healthy.

6.6.2.1. Ethical Considerations

- Legislation like the Genetic Information Non-discrimination Act (GINA) in the U.S. protects individuals from genetic discrimination in health insurance and employment. However, there is debate about whether such protections are adequate and how they apply in various contexts (e.g., life insurance).
- Ensuring equity and fairness in how genetic information is used and interpreted in decision-making processes.

6.6.3. Genetic Testing and Reproductive Choices

Prenatal and preimplantation genetic diagnosis (PGD), two technological advances in genetic testing, allow parents to screen for genetic disorders either before or during pregnancy. These technologies raise ethical questions even though they can help lower the prevalence of genetic diseases.

- **Eugenics Concerns**: Some fear that the use of genetic testing could lead to the practice of selecting embryos based on desired traits (such as intelligence or physical appearance), echoing historical eugenics movements that sought to "improve" the human race by selective breeding.
- **Disability Rights**: The use of genetic testing to avoid having children with disabilities raises concerns about the message it sends to individuals living with disabilities. It may reinforce negative perceptions of disability as something to be eliminated rather than embraced as part of human diversity.

6.6.3.1. Ethical Considerations

- The balance between reducing suffering through the prevention of genetic disorders and respecting the dignity and value of individuals with disabilities.
- Ensuring that reproductive decisions are made freely and not influenced by societal pressures or expectations regarding genetic "normalcy."

6.6.4. Equity in Access to Genetic Technologies

Not everyone has equal access to advances in genetics, particularly in personalized medicine and genetic therapies. This raises issues of fairness and social justice.

- **Socioeconomic Disparities**: Wealthier individuals or nations may have better access to cutting-edge genetic treatments and testing, potentially widening health disparities between rich and poor populations.
- **Global Inequality**: Genetic research often focuses on populations in wealthier countries, leading to a lack of representation of genetic variation in underrepresented populations from lower-income countries. This can result in treatments that are less effective for these populations.

6.6.4.1. Ethical Considerations

- Efforts should be made to ensure that genetic technologies are affordable and accessible to all populations, not just those with the means to pay for them.
- International collaborations should prioritize the inclusion of diverse populations in genetic research to ensure that the benefits of new discoveries are equitably shared.

6.6.5. Cultural and Social Impacts

Social perspectives on race, ethnicity, and identity can be influenced by research on human genetic variation. Both positive and negative effects may result from genetics' ability to support or refute beliefs about racial differences.

- **Race and Genetics**: Genetic research shows that race is a social construct with no clear biological basis, as there is more genetic variation within populations than between them. However, some groups may misuse genetic information to support racist ideologies.
- **Identity and Ancestry**: Genetic testing services that provide information on ancestry and heritage are growing in popularity, offering individuals a new way to explore their roots. While this can be empowering for some, it may also raise questions about identity, belonging, and the meaning of genetic ties.

6.6.5.1. Ethical Considerations

- Researchers and companies must be careful not to oversimplify or misinterpret genetic data in ways that reinforce harmful stereotypes or misunderstandings about race.
- The commercialization of ancestry testing raises concerns about the accuracy of results and the ethical implications of using genetic data to define personal identity.

Genetic variation in humans offers enormous opportunities as well as serious ethical problems. While technological and genetic advancements may lead to better health outcomes, privacy concerns, prejudice, inequality, and wider societal ramifications must be carefully considered. To guarantee that genetic science is applied responsibly and for the good of all people, ethical frameworks, laws, and public discourse are crucial.

Did you get it?

1. **Discuss about equity in access to genetic technologies**
2. **Focus on cultural and social impacts**

SUMMARY

- Human genetic variation is the genetic differences in and among populations. There may be multiple variants of any given gene in the human population (alleles), a situation called polymorphism.
- A genome is the complete set of DNA instructions found in every cell. DNA is made of four different chemicals (called nucleotides or bases), each represented by a different letter: adenine (A), thymine (T), cytosine (C) and guanine (G).
- The human population contains an immense amount of variation among genomes. A given reference human genome sequence is not intended to capture or represent that diversity; rather, it serves as an important data-analysis tool.
- Polymorphisms are genetic alterations that occur in more than 1% of the population. Most polymorphisms are what is known as single nucleotide polymorphism or SNP, meaning that the genetic alteration only affects the exchange of one nucleotide for another.
- Genetic variation is the raw material for evolution. It allows populations to adapt to changing environments through the process of natural selection.
- A single nucleotide polymorphism (SNP) is a difference in a single nucleotide between members of one species that occurs in at least 1% of the population.

MULTIPLE CHOICE QUESTIONS

1. **Which of the following is the most common type of genetic variation in humans?**
   a. Copy Number Variations (CNVs)
   b. Single Nucleotide Polymorphisms (SNPs)
   c. Insertions and Deletions (Indels)
   d. Structural Variations
2. **Which process involves the exchange of genetic material between homologous chromosomes during meiosis?**
   a. Mutation
   b. Genetic Drift
   c. Recombination
   d. Natural Selection
3. **Which of the following is a source of genetic variation?**
   a. Mutations
   b. Recombination
   c. Gene Flow
   d. All of the above

4. **A mutation that causes a change in a single nucleotide base pair is called a:**
   a. Copy Number Variation
   b. Single Nucleotide Polymorphism
   c. Insertion
   d. Gene Flow
5. **Which of the following conditions is an example of a single-gene disorder caused by genetic variation?**
   a. Diabetes
   b. Sickle Cell Anemia
   c. Hypertension
   d. Heart Disease

REVIEW QUESTIONS

1. What is human genetic variation, and why is it important for evolution and survival?
2. Explain how mutations contribute to genetic variation in human populations. Give an example of a beneficial mutation.
3. Describe the role of recombination during meiosis and how it contributes to genetic diversity.
4. What are single nucleotide polymorphisms (SNPs), and how do they influence traits in humans?
5. How does genetic variation affect the way individuals respond to diseases and medications? Provide an example.
6. Discuss the ethical concerns associated with using genetic information in personalized medicine and healthcare.

Answers to Multiple Questions

1. (b); 2. (c); 3. (d); 4. (b); 5. (b)

REFERENCES

1. Auton, A., Brooks, L. D., Durbin, R. M., Garrison, E. P., Kang, H. M., Korbel, J. O., et al. (2015). A global reference for human genetic variation. *Nature*, *526*(7571), 68–74.
2. Bamshad, M., & Wooding, S. P. (2003). Signatures of natural selection in the human genome. *Nature Reviews Genetics*, *4*(2), 99–111.
3. Cavalli-Sforza, L. L., & Feldman, M. W. (2003). The application of molecular genetic approaches to the study of human evolution. *Nature Genetics*, *33*(Suppl 3), 266–275.

4. Eiberg, H., et al. (2008). Blue eye color in humans may be caused by a perfectly associated founder mutation in a regulatory element located within the HERC2 gene inhibiting OCA2 expression. *Human Genetics*, *123*, 177–187.
5. Gabriel, S. B., Schaffner, S. F., Nguyen, H., Moore, J. M., Roy, J., Blumenstiel, B., Higgins, J., DeFelice, M., Lochner, A., Faggart, M., et al. (2002). The structure of haplotype blocks in the human genome. *Science*, *296*(5576), 2225–2229.
6. Harding, R. M., Healy, E., Ray, A. J., Ellis, N. S., Flanagan, N., Todd, C., Dixon, C., et al. (2000). Evidence for variable selective pressures at MC1R. *American Journal of Human Genetics*, *66*(4), 1351–1361.
7. Hernandez, R. D., et al. (2011). Classic selective sweeps were rare in recent human evolution. *Science*, *331*, 920–924.
8. Jorde, L. B., Bamshad, M., & Rogers, A. R. (1998). Using mitochondrial and nuclear DNA markers to reconstruct human evolution. *BioEssays*, *20*(2), 126–136.
9. Keita, S. O., Kittles, R. A., Royal, C. D., Bonney, G. E., Furbert-Harris, P., Dunston, G. M., & Rotimi, C. N. (2004). Conceptualizing human variation. *Nature Genetics*, *36*(11 Suppl), S17–S20.
10. Lamason, R. L., et al. (2005). SLC24A5, a putative cation exchanger, affects pigmentation in zebrafish and humans. *Science*, *310*, 1782–1786.
11. Li, H., & Durbin, R. (2011). Inference of human population history from individual whole-genome sequences. *Nature*, *475*, 493–496.
12. Moltke, I., et al. (2014). A common Greenlandic TBC1D4 variant confers muscle insulin resistance and type 2 diabetes. *Nature*, *512*, 190–193.
13. Race, Ethnicity, & Genetics Working Group. (2005). The use of racial, ethnic, and ancestral categories in human genetics research. *American Journal of Human Genetics*, *77*(4), 519–532.
14. Storey, J. D., Madeoy, J., Strout, J. L., Wurfel, M., Ronald, J., & Akey, J. M. (2007). Gene-expression variation within and among human populations. *American Journal of Human Genetics*, *80*(3), 502–509.
15. Templeton, A. R. (2018). *Human Population Genetics and Genomics* (pp. 445, 446). London.
16. Yi, X., et al. (2010). Sequencing of 50 human exomes reveals adaptation to high altitude. *Science*, *329*, 75–78.

CHAPTER 7

The Techniques of Molecular Genetics

LEARNING OBJECTIVES

After studying this chapter, you will be able to:

- Discuss the basic techniques used to identify, amplify, and clone genes
- Focus on construction and screening of DNA libraries
- Describe the molecular analysis of DNA, RNA, and protein
- Explain the molecular analysis of genes and chromosomes

INTRODUCTORY EXAMPLE

Fundamental Techniques of Molecular Genetics: The complex phenomena underlying living processes by identifying the genetic mechanisms that produce such processes. To access the causal structure of genetic mechanisms, biologists use sophisticated molecular techniques to manipulate the components of the mechanism and observe the resulting effects. Scientific knowledge in genetics therefore progresses in a distinctive way; progress is driven by the introduction and use of novel techniques.

UNIT INTRODUCTION

The study of biological trait variation and inheritance is known as genetics. As we've already mentioned, genetic research can be done without actually looking at DNA. In fact, some of the greatest geneticists relied on the study of phenotypes, inheritance patterns, and their ratios in carefully crafted crosses rather than having any particular expertise with DNA. By using a variety of methods to investigate and modify genetic material, molecular genetics provides deep understanding of inheritance and gene function. One important technique is Polymerase Chain Reaction (PCR), which allows for detailed analysis even with small samples by amplifying particular DNA sequences.

The exact nucleotide order is ascertained by DNA sequencing, including next-generation sequencing (NGS), which makes genetic variants and mutations visible. Targeted modifications of the genome are made possible by gene editing technologies like CRISPR-Cas9, which offer resources for researching the functions of individual genes and creating possible treatments. Gene expression studies quantify the levels of gene activity under various conditions using methods like RNA sequencing and quantitative PCR (qPCR). Additionally, DNA fragments are separated by size using gel electrophoresis, which makes genetic material easier to view and analyze. Together, these techniques improve our knowledge of genetic mechanisms, aid in the study of genomics and personalized medicine, and make it possible to create novel therapies for genetic illnesses. Molecular genetics continues to propel discoveries and applications that expand our understanding of biology and enhance human health thanks to these creative methods.

7.1. BASIC TECHNIQUES USED TO IDENTIFY, AMPLIFY, AND CLONE GENES

Learning Objectives

- What do you mean by DNA Cloning?
- Discuss the discovery of restriction endonucleases
- Explain the amplification of recombinant DNA molecules in cloning vectors
- Discuss the cloning large genes and segments of genomes in BACs, PACs, and YACs

A mammal's haploid genome has roughly 3×10^9 nucleotide pairs in it. The coding region of a gene represents one of a million sequences in the genome if the average gene's combined exons are 3000 nucleotide pairs long (many are longer). Even though genes make up less than 5% of the DNA in mammalian genomes, identifying a single gene is like trying to find a needle in a haystack. The majority of methods for analyzing genes and other DNA sequences demand that the sequence be readily available in sizable amounts in either pure or nearly pure form. The development of recombinant DNA and gene-cloning technologies has given molecular geneticists methods by which genes or other segments of large chromosomes can be isolated, replicated, and studied by nucleic acid sequencing techniques, electron microscopy, and other analytical techniques. However, the challenge remains in identifying the segment of a DNA molecule that carries a single gene and isolating enough of this sequence in pure form to permit molecular analyses of its structure and function. In fact, there are two different ways to amplify genes or other DNA sequences: one involves amplification taking place *in vivo,* while the other involves amplification occurring *in vitro.* Only when the short nucleotide sequences on either side of the target DNA sequence are known can the second method be applied.

In the second method, a unique (heat-stable) DNA polymerase is used to create short DNA strands that are complementary to the DNA sequences on either side of the gene or DNA sequence of interest. These strands are then used to start the amplification of the target DNA sequence *in vitro.* This process, known as the polymerase chain reaction (PCR), is a very effective gene-amplification instrument. After analysis and sequencing, the amplified products can be sequenced and, if needed, cloned into vectors for *in vivo* replication for further research. When a DNA sequence is amplified using PCR, it is often not necessary to clone the sequence through *in vivo* replication. Consequently, earlier *in vivo* amplification protocols have frequently been replaced with processes involving the PCR amplification of DNA sequences. PCR, however, is only applicable in situations where the nucleotide sequences flanking the target gene or DNA sequence are known.

To identify, amplify, and clone genes, scientists use a range of sophisticated techniques that allow for precise manipulation and study of genetic material.

Gene Identification:

- **Gene Mapping**: Techniques like linkage analysis and association studies help locate genes associated with specific traits or diseases. These methods use genetic markers and statistical analysis to pinpoint gene locations on chromosomes.
- **Bioinformatics**: Computational tools and databases, such as BLAST and Ensembl, are used to identify and predict gene sequences based on known genomic data.

Gene Amplification:

- **Polymerase Chain Reaction (PCR)**: PCR is a widely used method to amplify specific DNA sequences. By using short DNA primers that bind to the regions flanking the target gene, PCR replicates the gene of interest exponentially, making it easier to analyze and manipulate.
- **Quantitative PCR (qPCR)**: This variation of PCR allows for the quantification of gene expression levels in real-time, providing insights into gene activity and regulation.

Gene Cloning:

- **Molecular Cloning**: This involves inserting the gene of interest into a vector, such as a plasmid, which is then introduced into host cells (bacteria, yeast, or mammalian cells). The host cells replicate the vector, producing copies of the gene.
- **Recombinant DNA Technology**: Combining DNA fragments from different sources using restriction enzymes and ligases allows for the creation of recombinant DNA molecules. These can be used to produce proteins or study gene function.
- **CRISPR/Cas9**: Although primarily known for gene editing, CRISPR/Cas9 can also be used in conjunction with cloning methods to introduce genes into specific locations in the genome.

These techniques collectively enable researchers to isolate, study, and manipulate genes with high precision, facilitating advances in genetics, molecular biology, and biotechnology.

7.1.1. DNA Cloning

DNA cloning is a molecular biology technique which is used for the creation of exact copies or clones of a particular gene or DNA. DNA cloning is the process of making multiple copies of a particular segment of DNA. During this technique, the selected DNA fragment is inserted into a plasmid (the circular piece of DNA) using enzymes.

Restriction enzymes and DNA ligase are used in the process. The restriction enzymes are used to cut the DNA fragments at specific sequences and DNA ligase enzymes are used to join the nicks. The recombinant DNA thus produced is introduced into bacteria. These bacteria reproduce and produce an exact copy of the plasmid. These copies are known as clones.

7.1.1.1. DNA Cloning Steps

DNA Cloning takes place in the following steps:

- **Cutting and Pasting DNA:** Two types of enzymes are used in this method:
  - Restriction enzymes
  - DNA ligase

  The restriction enzymes cut the DNA at specific target sequences. The target gene is inserted into the cut site and is ligated by DNA ligase. This is known as a recombinant plasmid.
- **Bacterial Transformation and Selection:** The recombinant plasmid is introduced into bacteria such as E.coli. The bacteria are subjected to very high temperatures which compel them to take up the DNA. This process is known as transformation. The plasmid contains an antibiotic resistance gene which helps them to survive in the presence of antibiotics. The plasmid-containing bacteria are selected on a nutrient containing antibiotics. The transformed bacteria survive, while the ones without a plasmid die.
- **Protein Production:** The plasmid-containing bacteria are cultured, and the bacteria are provided with a chemical signal that helps them to target the protein. After protein production, the bacteria are split open to release it. The protein is purified, and the target protein is isolated from other contents of the cell.
- **Importance of DNA Cloning:** The DNA molecules produced through the cloning techniques are used for many purposes, which include:
  - DNA cloning can be used to make proteins such as insulin with biomedical techniques.
  - It is used to develop recombinant versions of the non-functional gene to understand the functioning of the normal gene. This is applied in gene therapies also.
  - It helps to analyze the effect of mutation on a particular gene.

Keyword

A plasmid is a small, circular DNA molecule that is separate from a cell's chromosomal DNA and can replicate independently.

7.1.2. The Discovery of Restriction Endonucleases

Restrictions endonucleases, whose name comes from the Greek word Éndon, which means "within," are a unique class of enzymes that are capable of cloning and sequencing nearly any gene or other interesting DNA sequence from any species. Endonucleases cut DNA molecules internally. While Type II restriction enzymes only cleave DNA molecules at particular nucleotide sequences known as restriction sites, many endonucleases randomly cut DNA. In contrast, restriction endonucleases are site-specific. Whatever the DNA's origin, type II restriction enzymes cleave it at these locations. Various microbes produce distinct restriction endonucleases that are specific to particular DNA nucleotide sequences. The first letter of the genus and the first two letters of the species that makes the enzyme are used to name the restriction endonucleases. If a strain is the only one that can produce an enzyme, the name of the strain is followed by a letter. The first restriction enzyme discovered from a strain of bacteria, II denotes the second, and so forth. Thus, Escherichia coli strain RY13 produces the restriction endonuclease EcoRI. There are restriction endonucleases that cleave DNA molecules at a variety of DNA sequences because hundreds of restriction enzymes have been identified and refined.

Table 7.1. Recognition Sequences and Cleavage Sites of Representative Restriction Endonucleases

Enzyme	Source	Recognition Sequence[a] and Cleavage Sites[b]		Products		Type of Ends Produced
EcoRI	*Escherichia coli* strain RY13	5′-GAA TTC -3′ 3′-CTT AAG-5′	Restriction digest →	5′-G 3′-CTTAA-5′ +	5′-AATTC-3′ G-5′	5′ Overhangs
HincII	*Haemophilus influenzae* strain R_c	5′-GTPy PuAC-3′ 3′-CAPu PyTG-5′	→	5′-GTPy-3′ 3′-CAPu-5′ +	5′-PuAC-3′ 3′-PyTG -5′	Blunt
HindIII	*Haemophilus influenzae* strain R_d	5′-AAG CTT-3′ 3′-TTC GAA-5′	→	5′-A 3′-TTCCGA-5′ +	5′-AGCTT-3′ A-5′	5′ Overhangs
HpaII	*Haemophilus parainfluenzae*	5′-CC GG-3′ 3′-GG CC-5′	→	5′-C 3′-GGC-5′ +	5′-CGG-3′ C-5′	5′ Overhangs
AluI	*Arthrobacter luteus*	5′-AG CT-3′ 3′-TC GA-5′	→	5′-AG-3′ 3′-TC-5′ +	5′-CT-3′ 3′-GA-5′	Blunt
PstI	*Providencia stuartii*	5′-CTG CAG-3′ 3′-GAC GTC-5′	→	5′-CTGCA-3′ 3′-G +	G-3′ 3′-ACGTC-5′	3′ Overhangs
ClaI	*Caryophanon latum*	5′-ATC GAT-3′ 3′-TAG CTA-5′	→	5′-AT 3′-TAGC-5′ +	5′-CGAT-3′ TA-5′	5′ Overhangs
SacI	*Streptomyces achromogenes*	5′-GAG CTC-3′ 3′-CTC GAG-5′	→	5′-GAGCT-3′ 3′-C +	C-3′ 3′-TCGAG-5′	3′ Overhangs
NotI	*Nocardia otitidis*	5′-GCGG CCGC-3′ 3′-CGCC GGCG-5′	→	5′-GC 3′-CGCCGG-5′ +	5′-GGCCGC-3′ CG-5′	5′ Overhangs

Source: https://plantlet.org/wp-content/uploads/2021/08/Capture-4.png.

[a]The axis of dyad symmetry in each palindromic recognition sequence is indicated by the red dot; the DNA sequences are the same reading in opposite directions from this point and switching the top and bottom strands to correct for their opposite polarity. Pu indicates that either purine (adenine or guanine) may be present at this position; Py indicates that either pyrimidine (thymine or cytosine) may be present.

[b]The position of each bond cleaved is indicated by an arrow. Note that with some restriction endonucleases the cuts are staggered (at different positions in the two complementary strands).

Hamilton Smith and Daniel Nathans discovered restriction endonucleases in 1970. Together with Werner Arber, whose groundbreaking work led to the discovery of restriction enzymes, they shared the 1986 Nobel Prize in Physiology or Medicine. The biological purpose of restriction endonucleases is to prevent foreign DNAs, such as DNA molecules from other species or viral DNAs, from "invading" the genetic material of bacteria. For this reason, restriction endonucleases are sometimes called prokaryotic immune systems.

An organism's own restriction endonucleases must shield all of its cleavage sites from cleavage; otherwise, the organism would destroy its own DNA and commit suicide. This methylation of one or more nucleotides in each nucleotide sequence that is recognized by the organism's own restriction endonuclease is often used to protect endogenous cleavage sites (Table 7.1). After replication, methylation proceeds quickly, assisted by the organism's production of site-specific methylases. A foreign DNA molecule will be broken up into a set number of fragments by each restriction endonuclease, which is determined by the number of restriction sites in the specific DNA molecule.

An interesting feature of restriction endonucleases is that they commonly recognize DNA sequences that are palindromes—that is, nucleotide-pair sequences that read the same forward or backward from a central axis of symmetry, as in the nonsense phrase

← →

AND MADAM DNA

In addition, a useful feature of many restriction nucleases is that they make staggered cuts; that is, they cleave the two strands of a double helix at different points (Figure 7.1). (Other restriction endonucleases cut both strands at the same place and produce blunt-ended fragments.) Because of the palindromic nature of the restriction sites, the staggered cuts produce segments of DNA with complementary single-stranded ends. For example, cleaving a DNA molecule of the following type:

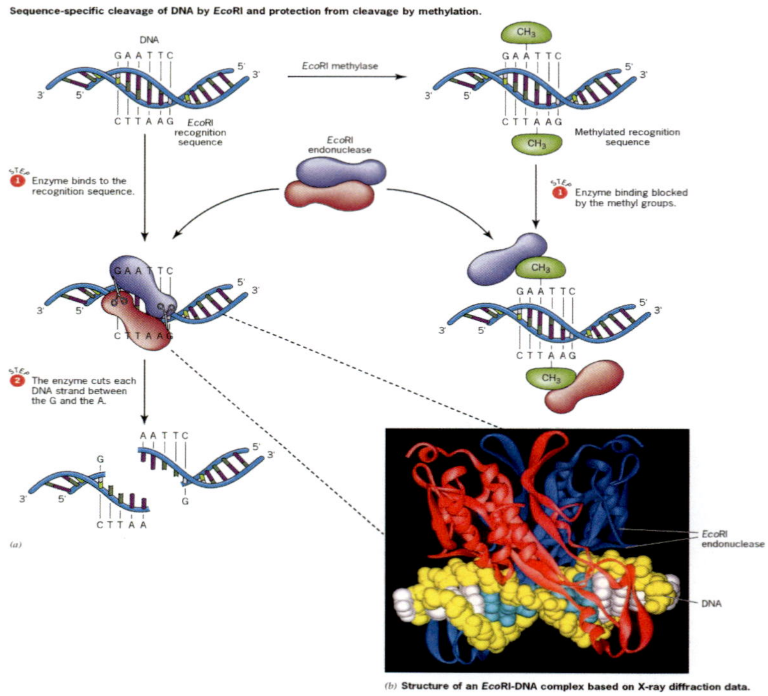

Figure 7.1. *The EcoRI restriction-modification system.*

Note: (a) Cleavage of the unmethylated EcoRI recognition sequence by EcoRI restriction endonuclease and protection of the recognition sequence from cleavage by methylation catalyzed by the EcoRI methylase. (b) Diagram of the structure of the EcoRI-DNA complex based on X-ray diffraction data. The two subunits of the EcoRI endonuclease are shown in red and blue, with the restriction endonuclease EcoRI will yield.

Source: https://www.google.co.in/books/edition/Principles_of_Genetics/NBB0CgAAQBAJ?hl=en&gbpv=1&dq=genetics+book&printsec=frontcover.

Because all the resulting DNA fragments will have complementary single stranded termini, they will hydrogen bond with each other and can be rejoined under the appropriate renaturation conditions by using the enzyme DNA ligase to re-form the missing phosphodiester linkages in each strand. Thus, DNA molecules can be cut into pieces, called restriction fragments, and the pieces can be joined together again with DNA ligase, almost at will.

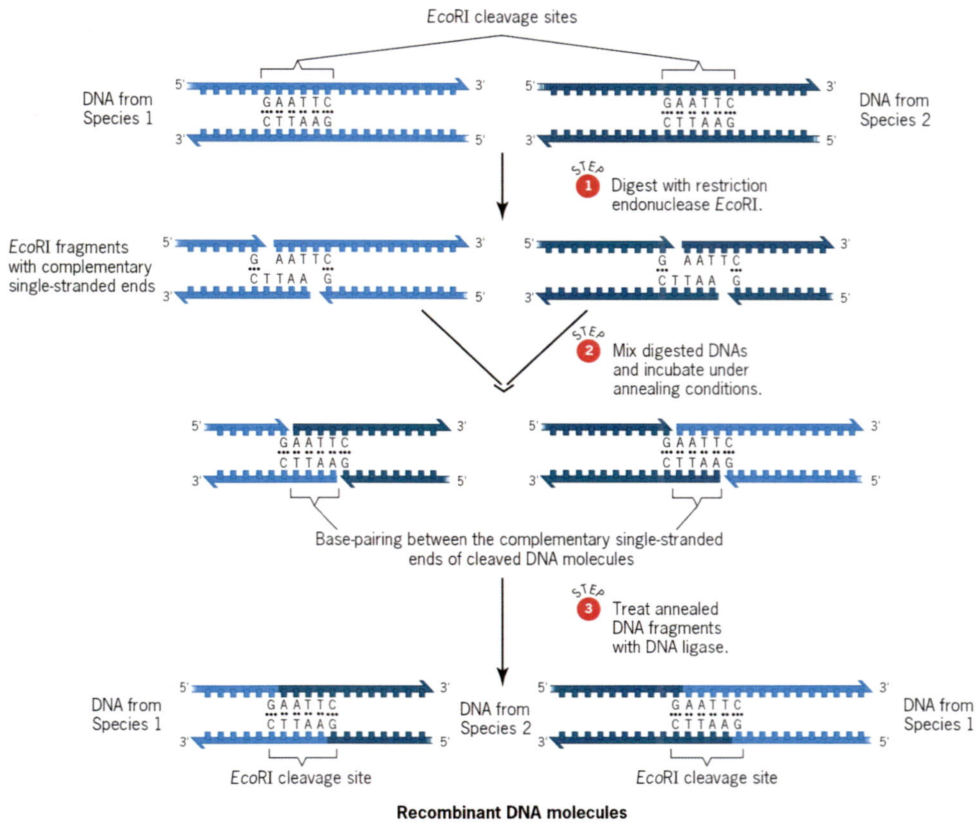

Figure 7.2. *The construction of recombinant DNA molecules in vitro. DNA molecules isolated from two different species are cleaved with a restriction enzyme, mixed under annealing conditions, and are covalently joined by treatment with DNA ligase. The DNA molecules can be obtained from any species—animal, plant, or microbe. The digestion of DNA with the restriction enzyme EcoRI produces the same complementary single stranded 5-AATT-3 ends regardless of the source of the DNA.*

Source: https://www.google.co.in/books/edition/Principles_of_Genetics/NBB0CgAAQBAJ?hl=en&gbpv=1&dq=genetics+book&printsec=frontcover.

7.1.3. The Production of Recombinant DNA Molecules In Vitro

Regardless of the DNA's origin, a restriction endonuclease catalyzes the cleavage of a particular sequence of nucleotide pairs. It's going to cut phage DNA, E. DNA from coli, maize, humans, or any other source as long as it has the recognized nucleotide sequence. Therefore, regardless of the source of DNA, the restriction endonuclease EcoRI will yield fragments with the same complementary single-stranded ends, 5-AATT-3, and two EcoRI fragments can covalently fuse together; that is, an EcoRI fragment from E can be joined to an EcoRI fragment from human DNA. coli DNA with the same ease as two EcoRI pieces from E. It is possible to combine two EcoRI fragments from human DNA or coli DNA. A recombinant DNA molecule is one that combines DNA fragments from two different sources, like the one displayed in Figure 7.2. The recombinant DNA technology that has revolutionized molecular biology over the last three decades is based on geneticists' ability to create such molecules at will.

Did you know?

The first recombinant DNA molecules were produced in Paul Berg's laboratory at Stanford University in 1972. Berg's research team constructed recombinant DNA molecules that contained phage lambda genes inserted into the small circular DNA molecule of simian virus 40 (SV40). In 1980, Berg was a co-recipient of the Nobel Prize in Chemistry as a result of this accomplishment.

Not long afterward, Stanley Cohen and associates from Stanford University introduced an EcoRI restriction fragment from a single DNA molecule into the distinct EcoRI restriction site of a plasmid that replicates itself. After inserting this recombinant plasmid into E. Similar to the original plasmid, it demonstrated independent replication in *E. coli* cells through transformation.

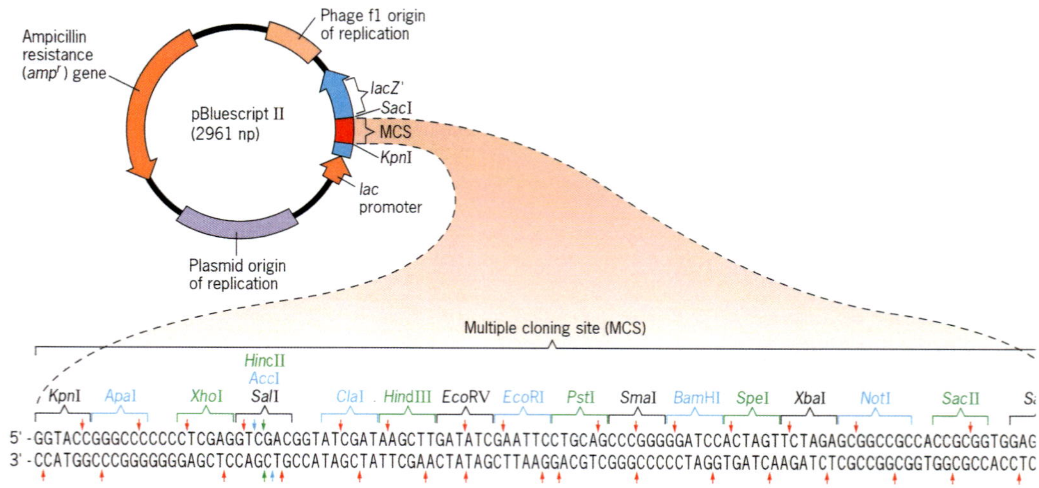

Figure 7.3. *The plasmid cloning vector.*

Note: (1) a plasmid origin of replication controlling double stranded DNA synthesis, (2) a phage f1 origin of replication controlling single-stranded DNA synthesis, (3) an ampicillin resistance gene (ampr) that serves as a dominant selectable marker, (4) the promoter for the lac genes and the promoter proximal segment (Z) of the lacZ gene, and (5) a poly-linker or multiple cloning site (MCS) containing a cluster of unique restriction enzyme cleavage sites (18 are shown). The MCS is located within the lacZ gene segment; therefore, when foreign DNA is inserted into the MCS, it disrupts LacZ function. The designators and brackets showing the locations of recognition sequences for the restriction enzymes are above the MCS DNA sequence. The cleavage sites are marked with red arrows except for AccI and HincII, where they are marked with blue and green arrows, respectively.

Source: https://media.cheggcdn.com/media/a00/a006764f-8664-4271-ba12-bb77e66a3545/phpYoOeet.png.

7.1.4. Amplification of Recombinant DNA Molecules in Cloning Vectors

Not only must recombinant DNA molecules be created, as illustrated in Figure 7.2, but they also need to be amplified, or produced in large quantities as clones, in order to be used in the diverse applications of recombinant DNA technologies. Making sure that one of the parental DNAs integrated into the recombinant DNA molecule is capable of self-replication allows us to achieve this. Essentially, the target gene or DNA sequence is transferred into a cloning vector that has been carefully selected.

A cloning vector consists of three essential components: (1) an origin of replication; (2) a dominant selectable marker gene, typically a gene that confers drug resistance to the host cell; and (3) at least one unique restriction endonuclease cleavage site, which is a cleavage site that is present only once in a region of the vector that does not disrupt the selectable marker gene or the origin of replication (Figure 7.3). The majority of commonly used cloning vectors are derived from plasmids or bacteriophage chromosomes. A group of distinct restriction sites known as a poly-linker or multiple cloning site is present in modern cloning vectors (Figure 7.3).

Keyword

Bacteria are microscopic living organisms that have only one cell.

A large number of cloning vectors are repurposed plasmids, which are circular DNA molecules that are extrachromosomal and found in bacteria. Plasmids can be as small as 1 kb (1 kilobase = 1000 base pairs) or as large as 200 kb, and many of them can replicate on their own. Antibiotic-resistant genes are also present in a lot of plasmids and make excellent selectable markers.

The fact that plasmid vectors can only accept relatively small foreign DNA inserts—maximum sizes of 10–15 kb—represents a limitation when using them. Scientists therefore looked for vectors that could continue to replicate in the presence of extremely large inserts. Table 7.2 contains a list of some of these vectors and the largest insert sizes that they will accept. After phage lambda vectors were widely utilized for a number of years, more complex vectors were created by fusing plasmid and viral components. Phagemids combine segments of plasmids with parts of phages, like M13. The cohesive ends of lambda, or cos sites, are contained in cosmids within plasmids. The origin of replication, centromere, and telomeres are the only essential components of yeast chromosomes found on yeast artificial chromosomes (YACs), which are linear minichromosomes with a selectable marker and multiple cloning sites. The essential elements of phage P1 chromosomes and bacterial fertility (F) factors are combined with multiple cloning sites and selectable marker genes to create bacterial artificial chromosomes (BACs) and P1 artificial chromosomes (PACs). Compared to plasmids and phage lambda cloning vectors, YACs, BACs, and PACs can accept substantially larger foreign DNA inserts (Table 7.2).

Table 7.2. Selected Cloning Vectors and Maximum Insert Sizes

Vector	Maximum Insert Size
Plasmids	15 kb
Phagemids	15 kb
Phage lambda	23 kb
Cosmids	44 kb
Bacterial artificial chromosomes (BACs)	300 kb
Phage P1 artificial chromosomes (PACs)	300 kb
Yeast artificial chromosomes (YACs)	600 kb

BlueScript (Figure 7.3) is a phagemid vector with a multiple cloning site (MCS) that contains many unique restriction enzyme cleavage sites, two distinct origins of replication, and a good selectable marker—a gene that makes the host bacterium resistant to ampicillin. The MCS is located within the 5 portion of the coding region of the lacZ gene, which encodes β-galactosidase, the enzyme that catalyzes the first step in the catabolism of lactose. When foreign DNA is inserted into one of the restriction sites in the MCS, it disrupts the function of the plasmid-encoded lacZ product. This inactivation of the amino-terminal segment of β-galactosidase provides a good visual test for determining whether or not the BlueScript plasmid in a cell contains a foreign DNA insert.

The basis for this visual test is as follows. The presence of β -galactosidase in cells can be monitored based on its ability to cleave the substrate 5-bromo-4-chloro-3-indolyl- β-D-galactoside (usually called X-gal) to galactose and 5-bromo-4- chloroindigo. X-gal is colorless; 5-bromo-4-chloroindigo is blue. Thus, cells containing active β -galactosidase produce blue colonies on agar medium containing X-gal, whereas cells lacking β-galactosidase activity produce white colonies on X-gal plates (Figure 7.4).

The β-galactosidase activity that yields the color indicator test for BlueScript vectors has a slightly more intricate molecular basis. The E gene lacZ. Given that coli is more than 3 kb long, the vector would be larger than desired if the entire gene were included in the plasmid. The lacZ gene is only partially present in the BlueScript vector. Only the amino-terminal region of β-galactosidase is encoded by this segment of the lacZ gene. Nonetheless, a distinct kind of complementation makes it possible to identify the existence of a functional copy of the lacZ gene segment. In cells with a specific lacZ mutant allele on the chromosome or an F plasmid, a functional copy of the lacZ gene segment on the BlueScript plasmid results in the production of polypeptides with combined β-galactosidase activity from the two faulty lacZ sequences. The lacZ ΔM15 mutant allele produces a Lac protein that is devoid of amino acids 11 through 14 from the amino terminus.

The mutant polypeptides cannot interact to form the active tetrameric form of the enzyme if these amino acids are absent. Tetramer formation by the ΔM15 deletion polypeptides is facilitated by the presence of the amino-terminal fragment (the first 147 amino acids) of the lacZ polypeptide encoded by the lacZ gene fragment on BlueScript plasmids. Because of the production of active β-galactosidase, the X-gal color test can be performed without enclosing the whole lacZ gene in the pBlueScript vector.

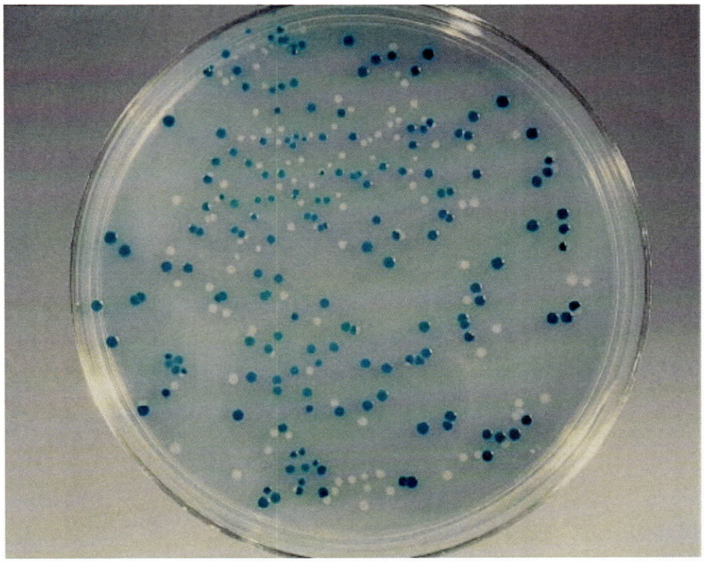

Figure 7.4. *Photograph illustrating the use of X-gal to identify E. coli colonies containing (blue) or lacking (white) β-galactosidase activity. In this case, the cells in the white colonies harbor blueScript plasmids with foreign DNA fragments inserted into the multiple cloning site, and the cells in the blue colonies contain blue-script plasmids with no insert.*

Source: https://biotechresources.com/wp-content/uploads/2019/06/bacillus-subtilis-in-lab.png.

7.1.5. Cloning Large Genes and Segments of Genomes in BACs, PACs, and YACs

Large eukaryotic genes are present. For instance, the human dystrophin gene, which carries a protein that connects muscle cell membranes to filaments, is more than 2,000 kb long. Using vectors that accept large foreign DNA inserts, such as BACs, PACs, and YACs, makes research on large genes and chromosomes much easier (*see* Table 7.2). Inserts between 300 and 600 kb can be used with these vectors. Compared to YACs, BACs and PACs are simpler to create and operate with. Furthermore, PACs and BACs are replicated in *E. coli* prefers plasmid carriers. Thus, in the study of large genes and genomes, like those of mammals and flowering plants, BAC and PAC vectors have largely replaced YAC vectors.

Negative selection against vectors without foreign DNA inserts is possible with PAC vectors, which have been created. The *Bacillus subtilis* sacB gene is present in these PAC vectors. The levansucrase enzyme, which is encoded by this gene, catalyzes the transfer of fructose groups to different types of carbohydrates. Selecting vectors containing inserts can be accomplished by inactivating the sacB gene by inserting foreign DNA into a BamHI restriction site. On a medium containing 5% sucrose, cells containing vectors with inserts can proliferate; cells containing vectors without inserts cannot. In the first hour of growth, cells with vectors lacking inserts lyse in the presence of 5% sucrose. Because of this, the vectors containing inserts that stop levansucrase activity within the sacB gene are present in every surviving cell. Figure 7.5 depicts one of these vectors' structural layout.

This shuttle vector, pJCPAC-Mam1, has two genes that help with vector replication in mammalian cells: the gene encoding Epstein-Barr virus's nuclear antigen 1 and the sacB gene, which permits positive selection of cells harboring vectors with inserts. Furthermore, the puromycin-resistance gene (pur r) has been incorporated to enable the selection of mammalian cells harboring the vector on puromycin-containing medium. BAC shuttle vectors that are comparable have also been created.

7.1.6. Amplification of DNA Sequences by the Polymerase Chain Reaction (PCR)

Many genomes, including the human genome, have nucleotide sequences available to us today that are either complete or almost complete. Researchers can isolate genes or other DNA sequences of interest without the need for cloning vectors or host cells thanks to the availability of these sequences in GenBank and other databases. The DNA sequence is amplified entirely *in vitro,* and it only takes a few hours to amp up the sequence by a millionfold or more. To use this procedure, one needs only be aware of the short nucleotide sequences that surround the sequence of interest. The polymerase chain reaction, or PCR for short, is used to amplify genes and other DNA sequences *in vitro.* PCR is a technique that primes the enzymatic amplification of the intervening segment of DNA in a test tube by using synthetic oligonucleotides complementary to known sequences flanking the sequence of interest. Kary Mullis invented the PCR process, which amplifies DNA sequences. For this work, he was awarded the 1993 Nobel Prize in Chemistry.

The PCR procedure involves three steps, each repeated many times (Figure 7.6). In step 1, the genomic DNA containing the sequence to be amplified is denatured by heating to 92–95°C for about 15 seconds. In step 2, the denatured DNA is annealed to an excess of the synthetic oligonucleotide primers by incubating them together at 50–60°C for 30 seconds.

The primer's base composition determines the optimal annealing temperature. The DNA segment between the sites complementary to the oligonucleotide primers is replicated in step 3 using DNA polymerase. The primer provides the free 3-OH required for covalent extension, and the denatured genomic DNA provides the required template function. In most cases, polymerization takes place for one to five minutes at 70 to 72°C. After being denatured, the products of the first replication cycle are annealed to oligonucleotide primers and replicated once more using DNA polymerase. The process is iterated numerous times until the required degree of complexity is attained. Keep in mind that amplification is geometric in nature. After one replication cycle, a single DNA double helix will produce two double helices, then 4 after 2, 8 after 3, 16 after 4, then 24 after 10, and so forth. The DNA sequence will have been duplicated more than a billion times after 30 amplification cycles.

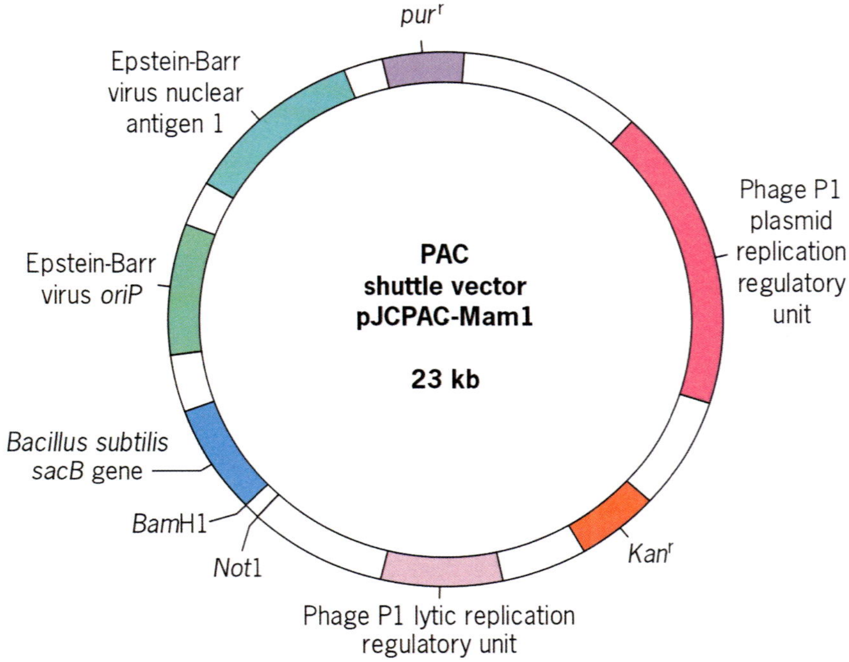

Figure 7.5. *Structure of the PAC mammalian shuttle vector pJCPAC-Mam1. The vector can replicate in either E. coli or mammalian cells. It can replicate in E. coli at low copy number under the control of the bacteriophage P1 plasmid replication unit or be amplified by inducing the phage P1 lytic replication unit (under the control of the lac inducible promoter). It can replicate in mammalian cells by using the origin of replication (oriP) and nuclear antigen 1 of the Epstein-Barr virus. Geneskanr and purr provide dominant selectable markers for use in E. coli and mammalian cells, respectively. The sacB gene (derived from Bacillus subtilis) is used for negative selection against vectors lacking DNA inserts. BamHI and NotI are cleavage sites for these two restriction endonucleases.*

Source: https://www.google.co.in/books/edition/Principles_of_Genetics/NBB0CgAAQBAJ?hl=en&gbpv=1&dq=genetics+book&printsec=frontcover.

Initially, DNA polymerase I of E was used for PCR as the replicase, *E. coli.* At the third step of each cycle, new enzyme had to be added because the denaturation step heat-inactivates this enzyme. *Thermus aquaticus*, a thermophilic bacterium, was found to contain a heat-stable DNA polymerase, which markedly improved DNA amplification by PCR. Taq polymerase, as it is commonly known, (*T. aquaticus* polymerase) doesn't stop working when the heat denaturation phase is underway. Consequently, there is no need to add polymerase following each denaturation cycle. Alternatively, oligonucleotide primers and excess Taq polymerase can be added at the beginning of the PCR process, and temperature changes can be used to carry out amplification cycles in a sequential manner. The amplification of particular DNA sequences by PCR is a reasonably easy process because PCR machines, also known as thermal cyclers, hold large quantities of samples and automatically adjust the temperature.

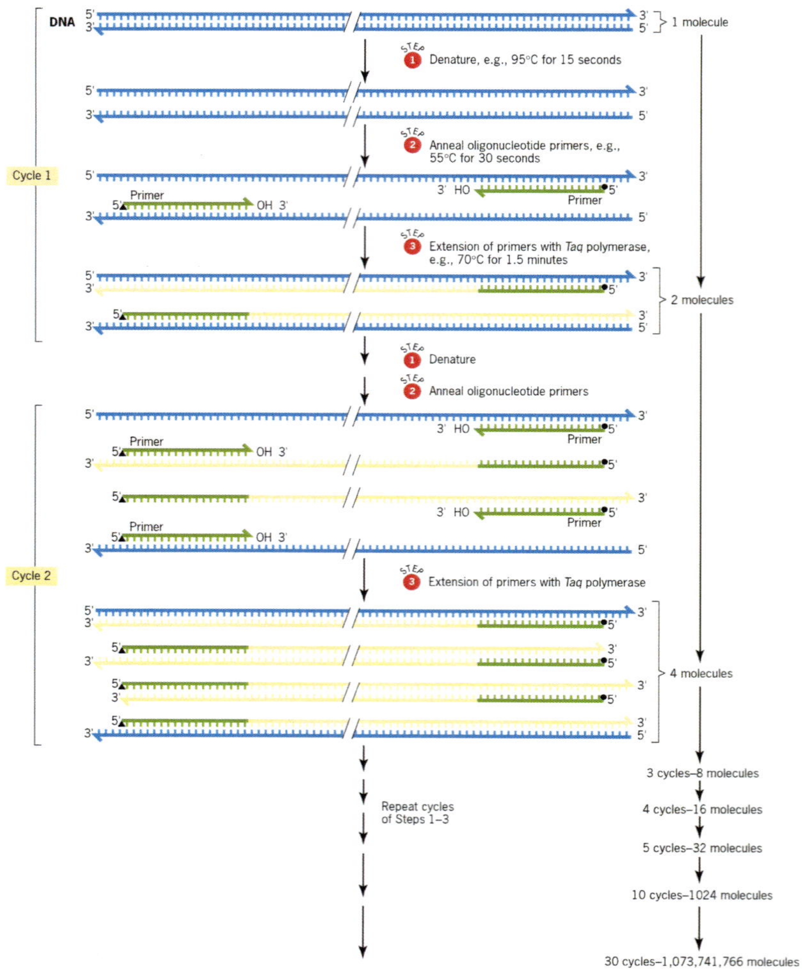

Figure 7.6. *The use of PCR to amplify DNA molecules in vitro.*

Note: Each cycle of amplification involves three steps: (1) denaturation of the genomic DNA being analyzed, (2) annealing of the denatured DNA to chemically synthesized oligonucleotide primers with sequences complementary to sites on opposite sides of the DNA region of interest; and (3) enzymatic replication of the region of interest by Taq polymerase.

Source: https://www.google.co.in/books/edition/Principles_of_Genetics/NBB0CgAAQBAJ?hl=en&gbpv=1&dq=genetics+book&printsec=frontcover.

The introduction of errors into the amplified DNA copies at low but significant frequencies is one drawback of PCR. Taq polymerase, in contrast to the majority of DNA polymerases, lacks an intrinsic 3 → 5 proofreading activity, which results in a higher-than-average frequency of replication errors. Like any other nucleotide in the DNA sequence, an incorrect nucleotide added during an early PCR cycle will be amplified. Heat-stable polymerases with 3 → 5 proofreading activity, such as Pfu from *Pyrococcus furiosus* or Tli from *Thermococcus litoralis*, are used in PCR when high fidelity is needed. Taq polymerase's inefficient amplification of lengthy DNA segments—more than a few 1000 nucleotide pairs—is one of its second drawbacks. Taq polymerase is substituted with the more processive Tfl polymerase from Thermus flavus when lengthy DNA segments need to be amplified. Up to 35 kb of DNA fragments can be amplified using Tfl polymerase. By PCR, fragments longer than 35 kb cannot be amplified effectively. For numerous applications that call for substantial quantities of a particular DNA sequence, PCR technologies offer workarounds.

When very small amounts of DNA are available, these procedures allow scientists to obtain definitive structural data on genes and DNA sequences. One significant use is in the diagnosis of hereditary human diseases, particularly in prenatal diagnosis situations when there is a scarcity of fetal DNA. A second significant use is in forensic cases where DNA extracted from minuscule tissue samples is used to identify specific individuals. DNA sequences are among the most reliable indicators of identity of any criteria. DNA sequences can be extracted from minuscule amounts of DNA found in a few drops of blood, semen, or even individual human hairs by employing PCR amplification. Thus, in legal cases involving ambiguous identities, PCR DNA profiling (also known as fingerprinting) experiments are crucial.

7.1.7. Treatment of Pituitary Dwarfism with Human Growth Hormone

In most ways, Kathy was a typical child—smart, joyful, energetic, and occasionally mischievous. True enough, Kathy's diminutive size was the only peculiar quality about her. Her birth defect was pituitary dwarfism, a condition caused by insufficient levels of human growth hormone (hGH). Kathy appeared destined to always be unusually diminutive. Kathy then started getting treatments with hGH made from bacteria when she was 10 years old. Throughout her first year of treatment, she gained five inches. Kathy's height reached the short end of the normal adult height distribution due to her continued hGH treatment during maturation. She would have continued to be abnormally small in stature without these treatments. Genetic engineering, which uses engineered or altered genes to produce desired products, produced the hGH that helped Kathy grow to almost normal size.

In the beginning, hGH was made in *E. coli* cells with a modified gene that is made up of synthetic bacterial regulatory elements fused to the hGH coding sequence. After being created *in vitro*, this chimeric gene was inserted into *E. coli* through transformation. hGH was produced in *E. coli* in 1985 and became the second genetically engineered pharmaceutical product to receive U.S. approval for use in humans from the Food and Drug Administration. The first product of this type was human insulin, which received approval in 1982. How do scientists create a gene in *E. coli* that produces human insulin or human growth hormone? They achieve this by fusing regulatory sequences that guarantee the expression of the human growth hormone or insulin gene in *E. coli* with the coding sequence of the respective gene. To enable expression, the gene must be inserted into living bacteria after it has been assembled in a test tube.

Thanks to the development of recombinant DNA technologies, molecular studies of genes and chromosomes have provided us with a great deal of information about the structure of genes. The cloning of particular genes is the first step in recombinant DNA techniques. A gene is cloned when it is isolated, inserted into a tiny self-replicating genetic element, like a plasmid or viral chromosome, and amplified when the plasmid or viral chromosome replicates in the proper host cell, which is typically an *E. coli* cell). Cloning vectors are tiny, self-replicating genetic elements that are used to clone genes. It is important to distinguish between the cloning of genes, which involves isolating and amplifying a specific gene, and the cloning of organisms, which involves creating an organism, like Dolly the lamb, from a single cell taken from an adult organism.

Remember

The synthesis of human proteins in bacteria was once thought to be science fiction. Human proteins are now regularly made in eukaryotic cells or bacteria that are cultured.

A particular gene's isolation and cloning is a difficult procedure. But once a gene has been cloned, it can undergo a wide range of modifications that enable studies into the connections between gene structure and function. A cloned gene is typically sequenced, meaning its nucleotide-pair sequence is found. The nucleotide sequence of a gene can be compared to thousands of other gene sequences kept in three sizable computer gene banks—one located in Germany, one in Japan, and one in the United States—if the gene's function is unknown. A gene's function can occasionally be inferred from its similarities to other genes whose functions are known. It is possible to predict the amino acid sequence of the polypeptide that a gene encodes using its nucleotide sequence and understanding of the genetic code. Next, amino acid sequences that might provide light on the polypeptide's function can be found by searching the predicted amino acid sequence of the protein.

Did you get it?

1. **Focus on amplification of DNA sequences by the polymerase chain reaction (PCR)**
2. **Learn about treatment of pituitary dwarfism with human growth hormone**

7.2. CONSTRUCTION AND SCREENING OF DNA LIBRARIES

Learning Objectives

- Discuss the construction of genomic libraries
- Focus on construction of cDNA libraries

The process of cloning an organism's genes typically begins with the creation of a genomic DNA library, which is a collection of DNA clones that together contain the entire genome. An organism's individual chromosomes can occasionally be separated using a method that divides chromosomes according to their size and DNA content. Chromosome-specific DNA libraries are then created using the extracted chromosomal DNAs. The search for a gene known to be located on a specific chromosome is made easier by the availability of chromosome-specific DNA libraries, especially for organisms with large genomes like humans. Libraries are created, amplified by replication, and then utilized to pinpoint specific genes or DNA sequences that the researcher is interested in.

An alternate method of gene cloning limits the search for genes to transcriptions of DNA sequences into copies of mRNA. Reverse transcriptase is an enzyme that is encoded by RNA retroviruses and it is responsible for catalyzing the synthesis of complementary DNA molecules to single-stranded RNA templates. Complementary DNAs (cDNAs) are the name given to these DNA molecules. DNA polymerases can be used to transform them into double-stranded cDNA molecules, which can then be cloned into plasmid carriers. Geneticists can create cDNA libraries that exclusively contain the coding regions of an organism's expressed genes by beginning with mRNA.

7.2.1. Construction of Genomic Libraries

In order to prepare genomic DNA libraries, an organism's total DNA is typically isolated, digested with a restriction endonuclease, and the restriction fragments are then inserted into the proper cloning vector. The restriction fragments can be directly ligated into vector DNA molecules that have been cut with the same restriction enzyme if the enzyme is used to make staggered cuts in DNA that result in complementary single-stranded ends (Figure 7.7). By using this method, the restriction endonuclease that was used to prepare the genomic DNA fragments for cloning can be used to cleave the foreign DNA inserts out of the vector DNA.

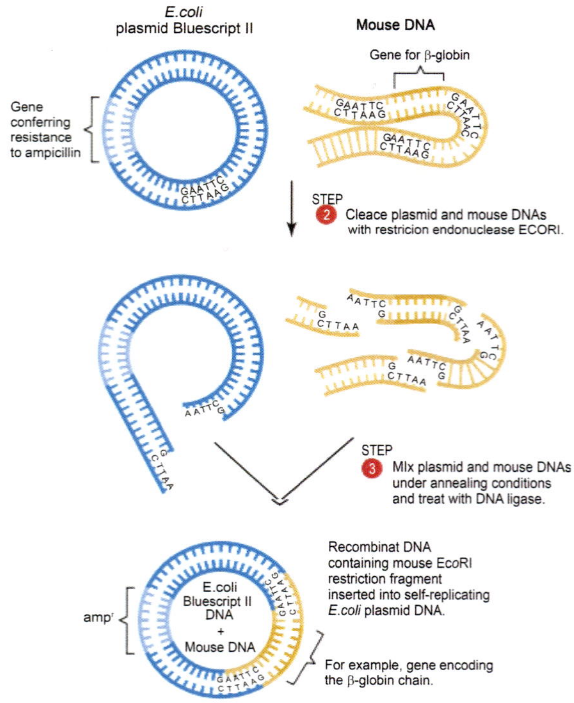

Figure 7.7. *Procedure used to clone DNA restriction fragments with complementary single-stranded ends.*

Source: https://www.google.co.in/books/edition/Principles_of_Genetics/NBB0CgAAQBAJ?hl=en&gbpv=1&dq=genetics+book&printsec=frontcover.

Keyword

A gene is a segment of DNA that is the basic unit of heredity and is passed from parents to offspring.

The recombinant DNA molecules must be inserted into host cells for *in vivo* replication amplification after the genomic DNA fragments have been ligated into vector DNA. The recipient cells that are susceptible to antibiotics are typically transformed, with the introduction of one recombinant DNA molecule per cell (for the majority of cells). As soon as *E. coli* is utilized, the bacteria must first be treated with chemicals or given a brief electrical pulse to make them soluble in DNA. After that, transformed cells are chosen by cultivating them in environments where the vector's selectable marker gene is necessary for cell growth. Nearly every DNA sequence found in the genome of interest can be found in a good genomic DNA library. Complete libraries for large genomes comprise hundreds of thousands of distinct recombinant clones.

7.2.2. Construction of cDNA Libraries

The majority of DNA sequences found in the extensive genomes of plants and higher animals do not encode proteins. Therefore, using complementary DNA (cDNA) libraries makes it easier to identify expressed DNA sequences. Reverse transcriptase can be primed to synthesize complementary DNA strands by using poly(T) oligomers, as the majority of mRNA molecules have three poly(A) tails (Figure 7.8). Then, ribonuclease H, DNA polymerase I, and DNA ligase work together to convert the RNA–DNA duplexes into double stranded DNA molecules. The RNA template strand is broken down by ribonuclease H, and the short RNA segments that are left over are used as primers for the synthesis of new DNA. DNA ligase seals the last single-strand breaks in the double-stranded DNA molecules while DNA polymerase I catalyzes the synthesis of the second DNA strand and substitutes DNA strands for RNA primers. These had two strands cDNAs can be inserted into plasmid or phage λ cloning vectors by adding complementary single-stranded tails to the cDNAs and vectors.

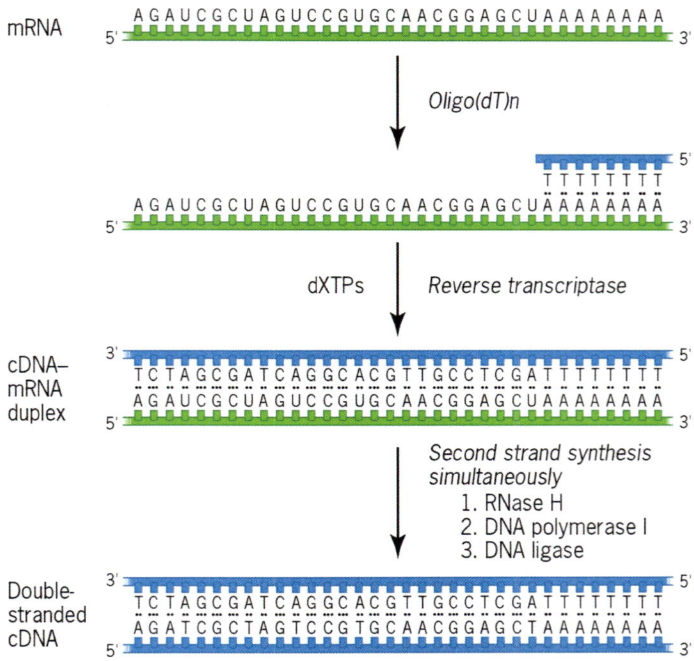

Figure 7.8. *The synthesis of double stranded cDNAs from mRNA molecules.*

Source: https://www.google.co.in/books/edition/Principles_of_Genetics/NBB0CgAAQBAJ?hl=en&gbpv=1&dq=genetics+book&printsec=frontcover.

7.2.3. Screening DNA Libraries for Genes of Interest

Higher plants and animals have very large genomes. For instance, there are 3 × 109 nucleotide pairs in the human genome. Therefore, identifying a single DNA sequence in a library that has a million or more distinct sequences is necessary when searching the genomic DNA or cDNA libraries of multicellular eukaryotes for a particular gene or other DNA sequence of interest. Genetic selection is the most effective screening method; it involves looking through a library of DNA sequences to find one that can give a mutant organism its wild-type phenotype back. More time-consuming molecular screens must be performed in situations where genetic selection is not an option. Typically, molecular screens use antibodies to detect gene products encoded by cDNA clones or DNA or RNA sequences as hybridization probes.

7.2.3.1. Genetic Selection

Genetic selection is the simplest method of locating an interesting clone. For instance, it is simple to clone the *Salmonella typhimurium* gene that confers penicillin resistance. A penR strain of *S. typhimurium* is used to create a genomic library. E. sensitive to penicillin. Coli cells are transformed using the library's recombinant DNA clones, and they are then plated on penicillin-containing medium. Penicillin will only allow the transformed cells carrying the penR gene to proliferate.

Remember

It is true that a large number of genes found in plants and animals have been found to be able to compensate for mutations in E. coli or yeast.

Genetic selection may be based on a gene's wild-type allele's capacity to give a mutant organism its normal phenotype when mutations in the gene of interest are present. Even though complementation screening is the name given to this sort of selection, its actual outcome is determined by the relative dominance of wild-type alleles over mutant alleles that encode inactive products. For instance, *E. coli* was used to clone the *S. cerevisiae* genes encoding histidine bio-synthesizing enzymes. *E. coli* histidine auxotrophs using clones of yeast cDNA and choosing transformed cells that could proliferate on a medium devoid of histidine.

There are restrictions on complementation screening. Introns are present in eukaryotic genes and need to be spliced out of gene transcripts before translation can occur. Due to *E. coli* cells lack the tools necessary to remove introns from eukaryotic genes, which makes it difficult to complement screening of eukaryotic clones in *E. coli.* Coli is limited to cDNAs, which have previously had their intron sequences removed. Furthermore, the proper transcription of the cloned gene in the new host is essential to the complementation screening process. Since eukaryotes have signals that regulate gene expression differently from prokaryotes, prokaryotic genes in prokaryotic organisms and eukaryotic genes in eukaryotic organisms are more likely to be compatible with the complementation approach. Because of this, *S. cerevisiae* is frequently used by researchers to screen eukaryotic DNA libraries using the complementation procedure.

7.2.3.2. Molecular Hybridization

Genes that are highly expressed in specialized cells were the first eukaryotic DNA sequences to be successfully cloned. These genes included the ovalbumin gene from chickens as well as the α and β-globin genes from mammals. Hemoglobin synthesis and storage require the high level of specialization found in red blood cells. Globin chains make up more than 90% of the protein molecules that red blood cells synthesize when they are at the peak of their biosynthesis. Similar to this, a primary byproduct of chicken oviduct cells is ovalbumin. This makes it simple to separate the RNA transcripts of the ovalbumin and globin genes from oviduct cells and reticulocytes, respectively. Using in situ colony or plaque hybridization, these RNA transcripts can be utilized to create radioactive cDNAs, which can then be utilized to screen genomic DNA libraries (Figure 7.9). Plaque hybridization is used with libraries built in phage lambda vectors, while colony hybridization is used with libraries built in plasmid and cosmid vectors. Although the two processes are nearly the same, in situ colony hybridization will be the main focus of this discussion.

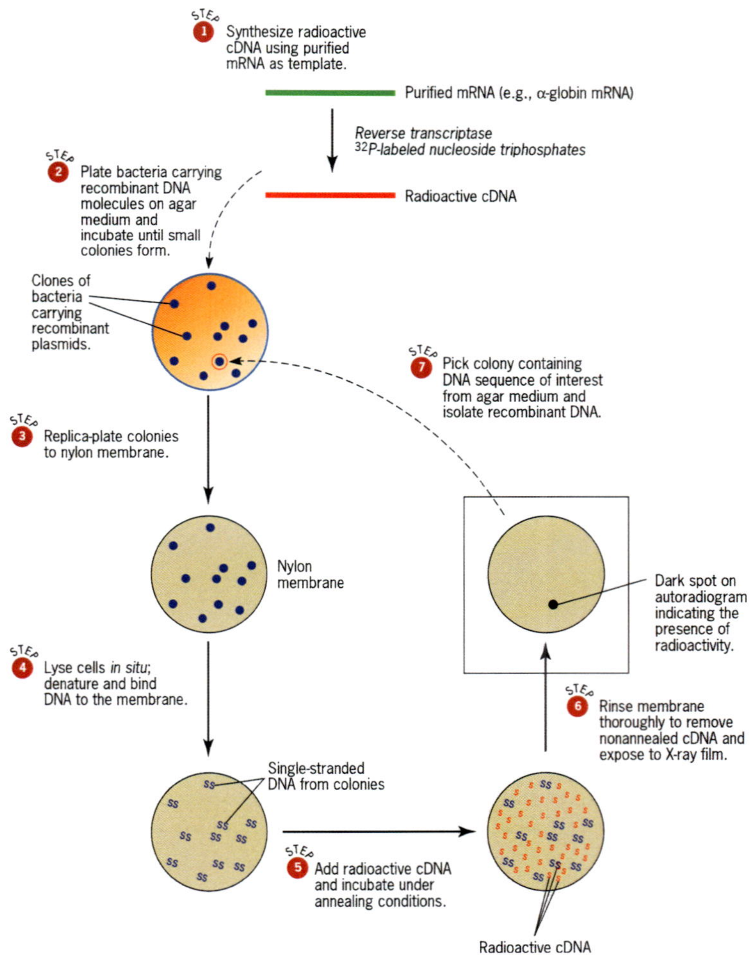

Figure 7.9. *Screening DNA libraries by colony hybridization. A radioactive cDNA is employed as a hybridization probe.*

Source: https://www.google.co.in/books/edition/Principles_of_Genetics/NBB0CgAAQBAJ?hl=en&gbpv=1&dq=genetics+book&printsec=frontcover.

The process of colony hybridization screening entails the transfer of transformed cell colonies onto nylon membranes, followed by autoradiography and hybridization with a radioactively labeled DNA or RNA probe (Figure 7.9). The labeled DNA or RNA is used to denatured DNA from colonies grown on the nylon membranes by hybridization. Prior to hybridization, the lysed cells' DNA is adhered to the membranes to prevent it from separating during the process's later stages. The membranes are exposed to X-ray film to detect the presence of radioactivity on them after time is allowed for hybridization between complementary strands of DNA. The membranes are then cleaned with buffered salt solutions to remove nonhybridized DNA. Radioactive spots on the autoradiographs will only appear in colonies that have DNA sequences complementary to the radioactive cDNA (Figure 7.9). On the original replicated plates, colonies containing the desired sequence are identified by the locations of the radioactive spots. These colonies are employed in the purification of DNA clones that carry the desired gene or DNA sequence. Work through Solve It: How Can You Clone a Specific NotI Restriction Fragment from the Orangutan Genome to see how well you understand the procedures used to prepare and screen genomic libraries.

Did you get it?

1. **How to screening DNA libraries for genes of interest?**
2. **What do you understand by molecular hybridization?**

7.3. THE MOLECULAR ANALYSIS OF DNA, RNA, AND PROTEIN

Learning Objectives

- How to analysis of DNAs by southern blot hybridizations?
- Focus on analysis of RNAs by northern blot hybridizations

Numerous novel methods for analyzing genes and gene products have emerged as a result of the development of recombinant DNA techniques. Just 25 years ago, questions that were virtually unanswerable could now be looked into rather easily. Almost any gene from any organism can be isolated and studied by geneticists; however, obtaining genes from large eukaryotic genomes can occasionally be a time-consuming and tedious process. It is possible to study a gene's expression in even the most complex organisms, like humans, once it has been cloned.

These and many other questions can now be routinely investigated using well-established methodologies. Is a specific gene expressed in the kidney, liver, bone cells, hair follicles, erythrocytes, or lymphocytes? Is this gene expressed throughout the organism's development or only during certain stages of development? Is a mutant allele of this gene similarly expressed, spatially and temporally, during development? Or does the mutant allele have an altered pattern of expression? If the latter, is this altered pattern of expression responsible for an inherited syndrome or disease? This text is not intended to be a comprehensive discussion of the methods used to study gene structure and function. Let's now examine some of the most significant techniques for studying the composition of genes (DNA), their transcripts (RNA), and their end products (mostly proteins).

7.3.1. Analysis of DNAs by Southern Blot Hybridizations

One effective method for separating macromolecules with varying sizes and charges is gel electrophoresis. DNA molecules almost exclusively separate on the basis of size or conformation in agarose and acrylamide gels because they have an almost constant charge per unit mass. Large molecules pass through molecular sieves made of acrylamide or agarose more slowly than small molecules. For molecules bigger than a few 100 nucleotides, agarose gels work better as sieves; for smaller DNA molecules, acrylamide gels work better. The separation of DNA restriction fragments by agarose gel electrophoresis is shown in Figure 7.10. Although the methods for separating RNA and protein molecules are essentially the same, they use slightly different approaches due to the distinctive characteristics of each class of macromolecule.

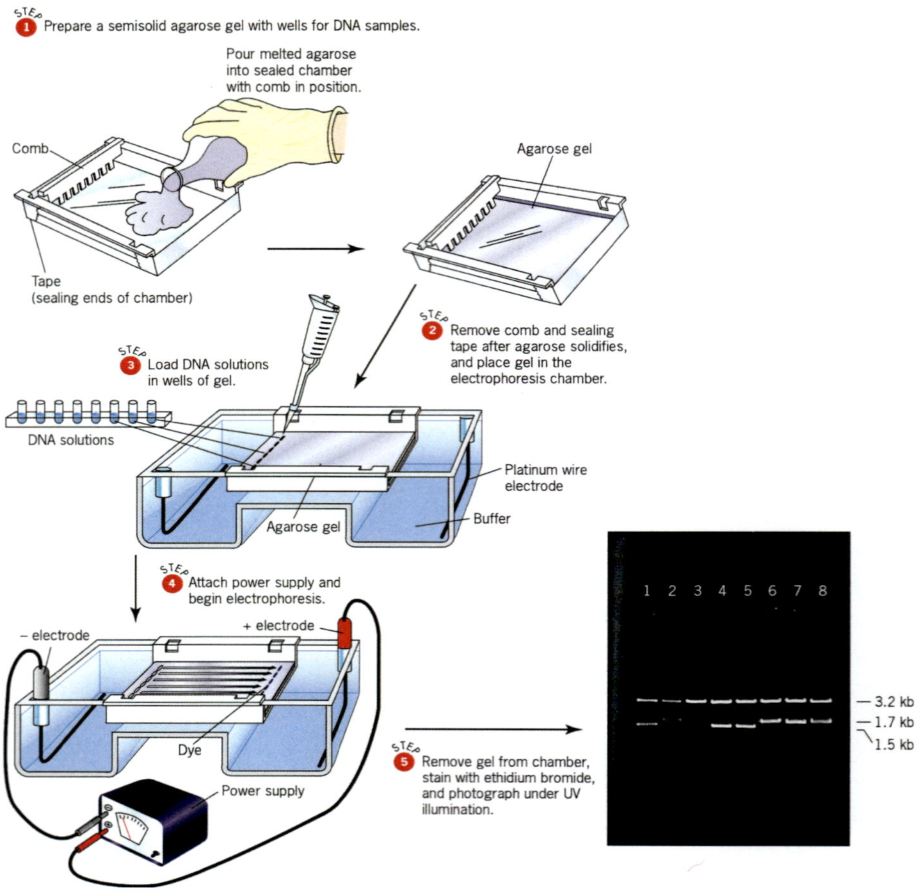

Figure 7.10. *The separation of DNA molecules by agarose gel electrophoresis. The DNAs are dissolved in loading buffer with density greater than that of the electrophoresis buffer so that DNA samples settle to the bottoms of the wells, rather than diffusing into the electrophoresis buffer. The loading buffer also contains a dye to monitor the rate of migration of molecules through the gel. Ethidium bromide binds to DNA and fluoresces when illuminated with ultraviolet light. In the photograph shown, lane 3 contained EcoRI-cut plasmid DNA; the other lanes contained EcoRI-cut plasmid DNAs carrying maize glutamine synthetase cDNA inserts.*

https://www.google.co.in/books/edition/Principles_of_Genetics/NBB0CgAAQBAJ?hl=en&gbpv=1&dq=genetics+book&printsec=frontcover.

The DNA molecules that have been separated by gel electrophoresis onto nitrocellulose or nylon membranes are the key component of this technique (Figure 7.11). The technique of transferring DNA to membranes is known as Southern blots, named after the scientist who invented it. The gel is submerged in an alkaline solution to denature the DNA before or during transfer. Following transfer, drying or UV light is used to immobilize the DNA on the membrane. Following the immobilized DNA on the membrane, a radioactive DNA probe containing the desired sequence is hybridized with it. Only DNA molecules with nucleotide sequences complementary to the probe's will cause the probe to hybridize. After washing the nonhybridized probe off the membrane, X-ray film is exposed to the cleaned membrane to look for radioactivity. The locations of DNA sequences that have hybridized with the probe are indicated by the dark bands on the developed film (Figure 7.12). One very helpful tool that has been developed is the ability to transfer DNA molecules separated by gel electrophoresis to nylon membranes for hybridization studies and other kinds of analyses.

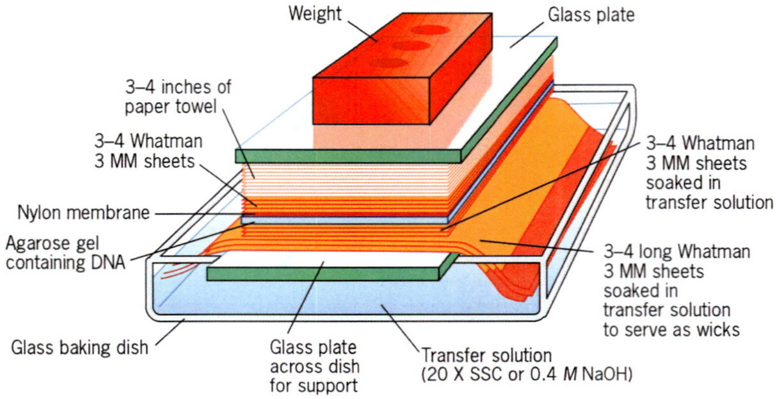

Figure 7.11. *Procedure used to transfer DNAs separated by gel electrophoresis to nylon membranes. The transfer solution carries the DNA from the gel to the membrane as the dry paper towels on top draw the salt solution from the reservoir through the gel to the towels. The DNA binds to the membrane on contact. The membrane with the DNA bound to it is dried and baked under vacuum to affix the DNA firmly prior to hybridization. SSC is a solution containing sodium chloride and sodium citrate.*

Source: https://www.google.co.in/books/edition/Principles_of_Genetics/NBB0CgAAQBAJ?hl=en&gbpv=1&dq=genetics+book&printsec=frontcover.

7.3.2. Analysis of RNAs by Northern Blot Hybridizations

We might anticipate that RNA molecules separated by agarose gel electrophoresis could be transferred and examined in a similar manner if DNA molecules can be transferred from agarose gels to nylon membranes for hybridization studies. In fact, genetics labs frequently use these kinds of RNA transfers. Since the process of RNA blots is similar to that of Southern blotting, but involves the separation and membrane transfer of RNA molecules, the technique is known as a "northern blot." This terminology has been expanded to include the process known as western blotting, which is the transfer of proteins from gels to membranes.

The process for Northern Blot transfers is nearly the same as that for Southern Blot transfers (Figure 7.11). RNases can, however, degrade RNA molecules very easily. Therefore, it's important to take precautions to avoid contaminating materials with these incredibly stable enzymes. Moreover, in order to separate RNA molecules based on size during electrophoresis, the majority of them must be maintained denatured due to their significant secondary structure. Formaldehyde or another chemical denaturant is added to the electrophoresis buffer to achieve denaturation. Similar to a Southern blot, the RNA blot is hybridized to either DNA or RNA probes after being transferred to the proper membrane. When studying gene expression, northern blot hybridizations (Figure 7.13) are very useful.

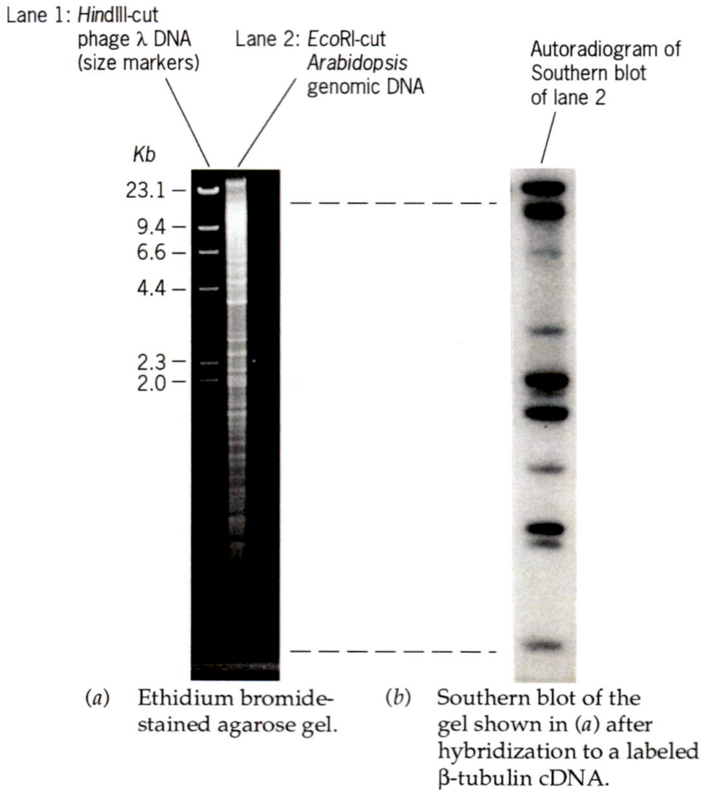

Figure 7.12. *Identification of genomic restriction fragments harboring specific DNA sequences by the Southern blot hybridization procedure.*

Note: (a) Photograph of an ethidium bromide-stained agarose gel containing phage DNA digested with HindIII (left lane), and Arabidopsis thaliana DNA digested with EcoRI (right lane). The DNA digest provides size markers. The A. thaliana DNA digest was transferred to a nylon membrane by the Southern procedure (Figure 7.11) and hybridized to a radioactive DNA fragment of a cloned β-tubulin gene. The resulting Southern blot is shown in (b); nine different EcoRI fragments hybridized with the β-tubulin probe.

Source: https://www.google.co.in/books/edition/Principles_of_Genetics/NBB0CgAAQBAJ?hl=en&gbpv=1&dq=genetics+book&printsec=frontcover.

They can be used to ascertain the location and timing of a gene's expression. But it's important to keep in mind that northern blot hybridizations only quantify the amount of RNA transcripts accumulated. They don't explain why the accumulation that has been seen has happened. Variations in the rate of transcription or the rate of transcript degradation may be the cause of changes in transcript levels. To separate these possibilities, more advanced processes need to be employed.

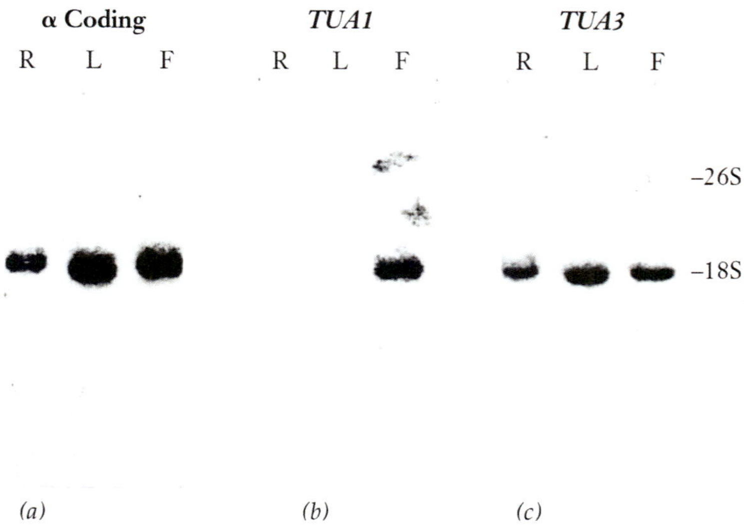

Figure 7.13. *Typical northern blot hybridization data. Total RNAs were isolated from roots (R), leaves (L), and flowers (F) of A. thaliana plants, separated by agarose gel electrophoresis, and then transferred to nylon membranes.*

Note: The autoradiogram shown in (a) is of a blot that was hybridized to a radioactive probe containing an α-tubulin coding sequence. This probe hybridizes to the transcripts of all six α -tubulin genes in A. thaliana. The autoradiograms shown in (b) and (c) are of RNA blots that were hybridized to DNA probes specific for the α1- and α3-tubulin genes (TUA1 and TUA3, respectively). The results show that the α3-tubulin transcript is present in all organs analyzed, whereas the α1-tubulin transcript is present only in flowers. The 18S and 26S ribosomal RNAs provide size markers. Their positions were determined from a photograph of the ethidium-bromide stained gel prior to transfer of the RNAs to the nylon membrane.

Source: https://www.google.co.in/books/edition/Principles_of_Genetics/NBB0CgAAQBAJ?hl=en&gbpv=1&dq=genetics+book&printsec=frontcover.

7.3.3. Analysis of RNAs by Reverse Transcriptase-PCR (RT-PCR)

The synthesis of complementary DNA strands to RNA templates is catalyzed by the enzyme reverse transcriptase. It can be applied to the *in vitro* synthesis of complementary DNA strands to RNA template strands. The resulting DNA strands can then be transformed into double-stranded DNA using a variety of techniques (for instance, as shown in Figure 7.8), such as the heat-stable Taq DNA polymerase and the addition of a second primer. Standard PCR can then be used to amplify the resultant DNA molecules. Gene-specific primers (sequences complementary to the RNA molecule of interest) or an oligo(dT) primer that will anneal to the 3-poly(A) tails of all mRNAs can be used to synthesize the first strand of DNA, which is commonly referred to as a cDNA because it is complementary to the mRNA under study. Primers that are specific to a gene are typically selected to anneal to sequences found in the 3′ noncoding regions of mRNAs. Such primers can be used in RT-PCR to amplify a specific gene transcript, as shown in Figure 7.14. Gel electrophoresis is utilized to analyze the results of these amplifications.

Numerous RT-PCR procedure modifications have been developed, with a primary focus on increasing its quantitative nature. To ascertain the correlation between RNA input and DNA output, for instance, known concentrations of the RNA under investigation can be examined. An investigator can use the amount of DNA produced by an experimental sample to calculate the amount of RNA that was initially present in that sample by using this relationship.

7.3.4. Analysis of Proteins by Western Blot Techniques

One crucial method for the separation and analysis of proteins is polyacrylamide gel electrophoresis. Individual polypeptides are separated by electrophoresis in the presence of the detergent sodium dodecyl sulfate (SDS), which denatures the proteins, since many functional proteins are made up of two or more subunits.

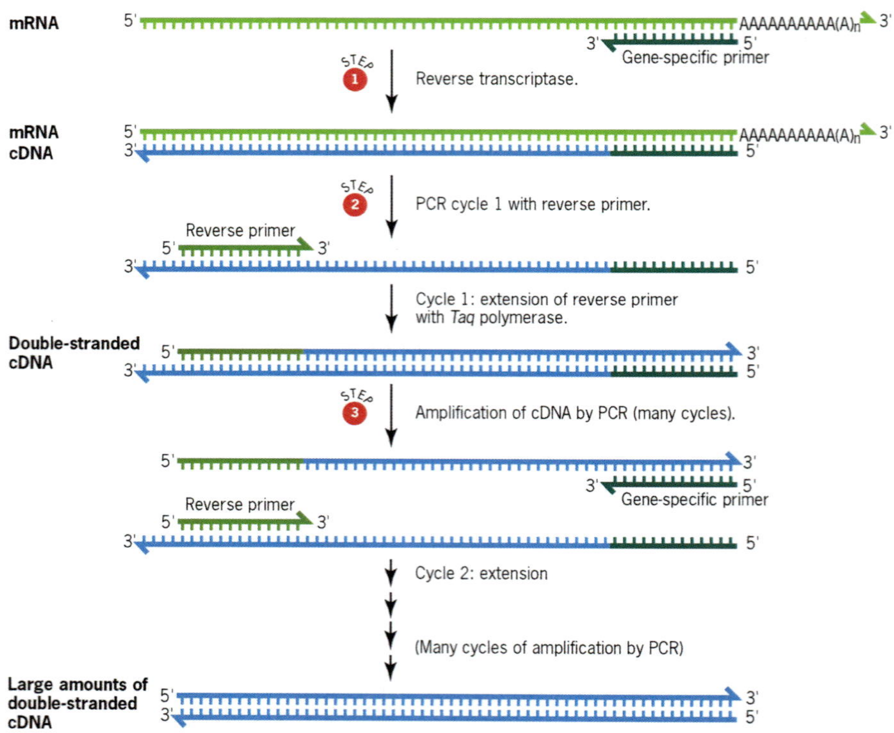

Figure 7.14. *Detection and amplification of RNAs by reverse transcriptase PCR (RT-PCR). Specific gene transcripts are amplified by first using reverse transcriptase to synthesize a single-stranded DNA that is complementary to the mRNA of interest. The synthesis is initiated with a gene-specific oligonucleotide primer (a primer that will only anneal to the mRNA of interest). The complementary DNA strand is then synthesized by using a reverse primer and Taq polymerase. Large quantities of double-stranded cDNA are subsequently synthesized by standard PCR reactions in the presence of both the gene-specific and reverse PCR primers.*

Source: https://www.google.co.in/books/edition/Principles_of_Genetics/NBB0CgAAQBAJ?hl=en&gbpv=1&dq=genetics+book&printsec=frontcover.

Silver or Coomassie blue staining is used to identify the proteins following electrophoresis. But the separated polypeptides can also be moved from the gel to a nitrocellulose membrane, where antibodies can be used to identify individual proteins. Western blotting, the process of transferring proteins from acrylamide gels to nitrocellulose membranes, is carried out by applying an electric current to the gel's surface to transfer the proteins. After transfer, the membrane containing the immobilized proteins is placed in a solution containing an antibody to the protein in order to identify a particular protein of interest. After washing the membrane with non-bound antibodies, the membrane is placed in a solution containing a secondary antibody to detect the presence of the first (primary) antibody. The group of proteins that make up all antibodies, known as immunoglobulins, are generally reacted with by this secondary antibody. The secondary antibody is coupled to an enzyme that, when the appropriate substrate is added, yields a visible product or to a radioactive isotope that allows for autoradiography.

Did you get it?

1. **Discuss the analysis of RNAs by reverse transcriptase-PCR (RT-PCR)**
2. **Explain the analysis of proteins by western blot techniques**

7.4. THE MOLECULAR ANALYSIS OF GENES AND CHROMOSOMES

Learning Objectives

- Discuss the physical maps of DNA molecules based on restriction enzyme cleavage sites
- Focus on nucleotide sequences of genes and chromosomes

Geneticists can ascertain the structure of individual genes, chromosomes, and entire genomes through the use of recombinant DNA techniques. Molecular geneticists have, in fact, created intricate physical and genetic maps of many organisms' genomes.

The complete nucleotide sequences of thousands of viruses, bacteria, mitochondria, chloroplasts, and various eukaryotic organisms have already been identified. Nucleotide sequences are the ultimate physical map of a genetic element. A "nearly complete" sequence of the human genome was released by the International Human Genome Sequencing Consortium in October 2004. That sequence covered 99% of the gene-rich chromatin in the human genome with only 341 gaps. The creation of restriction enzyme cleavage site maps of genes and chromosomes, as well as the identification of DNA sequences.

7.4.1. Physical Maps of DNA Molecules Based on Restriction Enzyme Cleavage Sites

According to Table 7.1, the majority of restriction endonucleases cleave DNA molecules only at specific sites. Because of this, they can be used to create physical chromosomal maps that are extremely helpful to researchers in helping to isolate DNA segments that carry genes or other sequences of interest. Polyacrylamide or agarose gel electrophoresis can be used to measure the sizes of the restriction fragments (*see* Figure 7.10). DNA has one phosphate group per nucleotide due to its nucleotide subunit structure, which gives DNA an almost constant charge per unit of mass. Because the rate of migration is inversely proportional to length, it is possible to accurately estimate the lengths of DNA fragments by measuring their migration rates during electrophoresis.

Figure 7.15 shows the process that is used to map the restriction enzyme cleavage sites. A set of DNA markers with known sizes is used to estimate the sizes of DNA restriction fragments. A collection of DNA molecules with length differences of 1000 nucleotide pairs are employed as size markers in Figure 7.15. Imagine a DNA molecule that is roughly six kilobases (6 kb) long. EcoRI cuts the 6-kb DNA molecule into two pieces, each measuring 4,000 and 2,000 nucleotide pairs. Figure 7.15b illustrates the potential locations of the molecule's lone EcoRI cleavage site. HindIII cleaves a single DNA molecule into two fragments, each measuring 5,000 or 1000 nucleotide pairs.

Figure 7.15c depicts the potential locations of the single HindIII cleavage site. Keep in mind that no conclusions regarding the relative locations of the EcoRI and HindIII cleavage sites can be drawn at this point in the analysis. Both of the two EcoRI restriction fragments contain the HindIII cleavage site. Following simultaneous digestion of the molecule by EcoRI and HindIII, three fragments with nucleotide pair sizes of 3,000, 2,000, and 1000 are created. The locations of the two cleavage sites on the molecule are determined by this outcome. The HindIII cleavage site must be at the other end of the molecule from the EcoRI cleavage site since the 2,000-nucleotide-pair EcoRI restriction fragment is still present and has not been cut by HindIII (Figure 7.15d). This kind of analysis can be expanded to create more comprehensive maps of restriction sites by utilizing multiple restriction enzymes. These restriction maps differ from genetic maps in that they represent actual physical distances along the DNA molecule, which is an important feature.

Keyword

A nucleotide sequence is the order of bases in a DNA or RNA molecule.

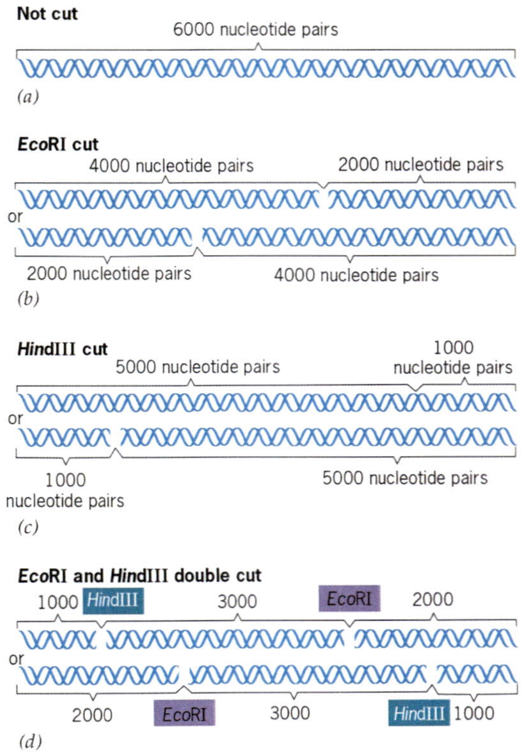

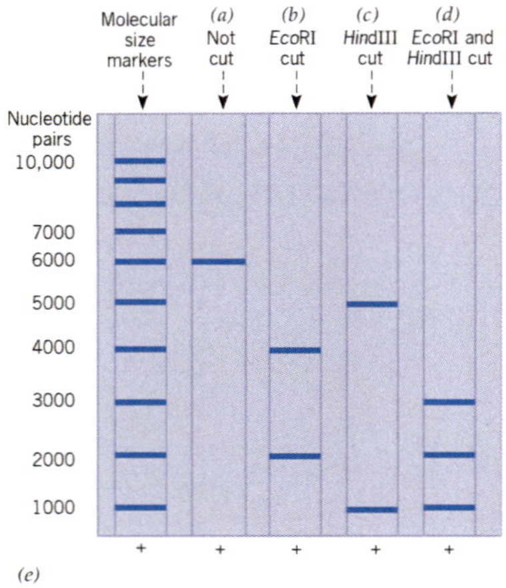

Figure 7.15. *Procedure used to map restriction enzyme cleavage sites in DNA molecules.*

Note: (a–d) Structures of the DNA molecule or of restriction fragments of the molecule either (a) uncut or cut with (b) EcoRI, (c) HindIII, or (d) EcoRI and HindIII. (e) The separation of these DNA molecules and fragments by agarose gel electrophoresis. The left lane on the gel contains a set of molecular size markers, a set of DNA molecules of size 1000 nucleotide pairs and multiples thereof.

Source: https://www.google.co.in/books/edition/Principles_of_Genetics/NBB0CgAAQBAJ?hl=en&gbpv=1&dq=genetics+book&printsec=frontcover.

Physical maps of whole genomes can be created by combining computer-assisted restriction mapping with other molecular techniques. An important worm for research on the genetic regulation of development, *Caenorhabditis elegans*, is the first multicellular eukaryote for which this has been achieved. Additionally, the actual map of the C. The genome and genetic map of *C. elegans* have been correlated. As a result, whenever a novel and intriguing mutation in *C. elegans*, whose location on the genetic map frequently enables the extraction of wild-type gene clones from a sizable international C. clone bank of elegans.

7.4.2. Nucleotide Sequences of Genes and Chromosomes

The nucleotide-pair sequence, which includes a list of all nucleotide-pair modifications that affect a gene's or chromosome's function, is the definitive physical map of a particular gene or chromosome. Before 1975, the idea of trying to sequence a whole chromosome was almost unthinkable; at best, it would have taken years of work and labor. However, the entire 5,386-nucleotide phage X174 chromosome had been sequenced by late 1976. Sequencing is now a standard laboratory process. 2,442 viruses, 1,372 bacteria, 93 archaea, and 40 eukaryotes have complete nucleotide sequences known, and genome sequencing projects are underway for another 3670 bacteria, 84 archaea, and 612 eukaryotes. Furthermore, 99% of the human genome's euchromatin sequence is known.

Four significant advancements led to our initial capacity to sequence almost any DNA molecule. The most significant development was the identification of restriction enzymes and their application in the preparation of uniform samples of particular chromosomal segments. The development of gel electrophoresis techniques to the point where DNA chains with single nucleotide length differences could be resolved represented another significant advancement. Techniques for cloning genes were also crucial because they made it easier to prepare large amounts of a specific DNA molecule. Ultimately, effective methods for determining the nucleotide sequences of DNA molecules were developed by researchers.

DNA sequencing procedures rely on the production of a population of DNA fragments that terminate at every possible position (every subsequent nucleotide) at one end and share one end (every nucleotide). The sequencing primer's 5-terminus serves as the common end. The site of chain extension by DNA polymerase is the free OOH present in the primer's 3-terminus. Fragments with variable three ends—ends at each potential nucleotide position along the DNA strand—are created by chain extension. Following that, these fragments are divided by chain length using polyacrylamide gel electrophoresis.

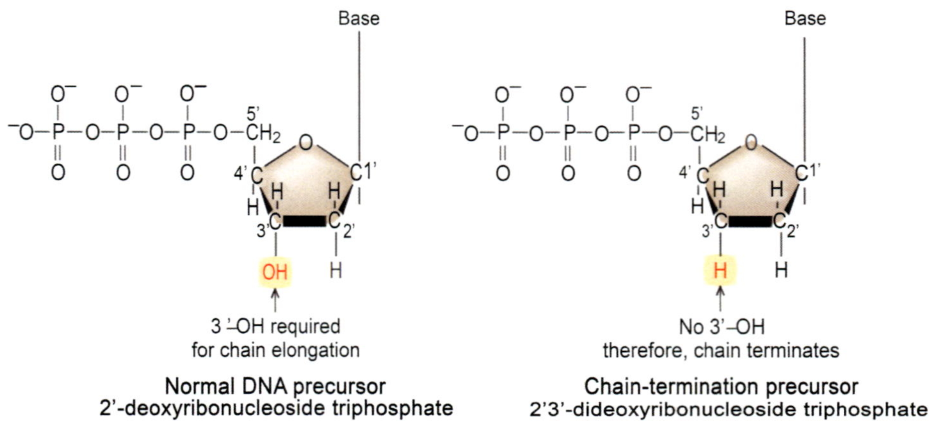

Figure 7.16. *Comparison of the structures of the normal DNA precursor 2-deoxyribonucleoside triphosphate and the chain-terminator 2,3-dideoxyribonucleoside triphosphate used in DNA sequencing reactions.*

Source: https://www.google.co.in/books/edition/Principles_of_Genetics/NBB0CgAAQBAJ?hl=en&gbpv=1&dq=genetics+book&printsec=frontcover.

These days, automated DNA sequencing machines are used for all DNA sequencing. The DNA sequencing protocol that Frederick Sanger and associates published in 1977 was initially used by sequencing machines. For this work, Sanger was awarded a share of the 1980 Nobel Prize in Chemistry. He was also awarded the 1958 Nobel Prize in Chemistry for deciphering the insulin's amino acid sequence.

The Sanger method is being replaced by faster and more advanced DNA sequencing techniques nowadays. The Sanger method creates populations of DNA fragments that terminate at As, Gs, Cs, and Ts, respectively, by using *in vitro* DNA synthesis in the presence of particular chain-terminators. ddXTPs, or 2,3-dideoxyribonucleoside triphosphates, are the most commonly used chain-terminators in the Sanger sequencing protocol (Figure 7.16). Remember that a free 3OOH on the DNA primer strand is a necessary condition for DNA polymerases. Due to the fact that 2,3-dideoxynucleotides lack 3-OH, adding one to the end of a chain will prevent further chain extension. A population of nascent fragments with chains with three termini at every possible position will be generated by using (1) 2,3-dideoxythymidine triphosphate (ddTTP), (2) 2,3-dideoxycytidine triphosphate (ddCTP), (3) 2,3-dideoxyadenosine triphosphate (ddATP), and (4) 2,3-dideoxyguanosine triphosphate (ddGTP) as chain terminators in a DNA synthesis reaction. Each of these compounds is labeled with a dye that fluoresces a different color.

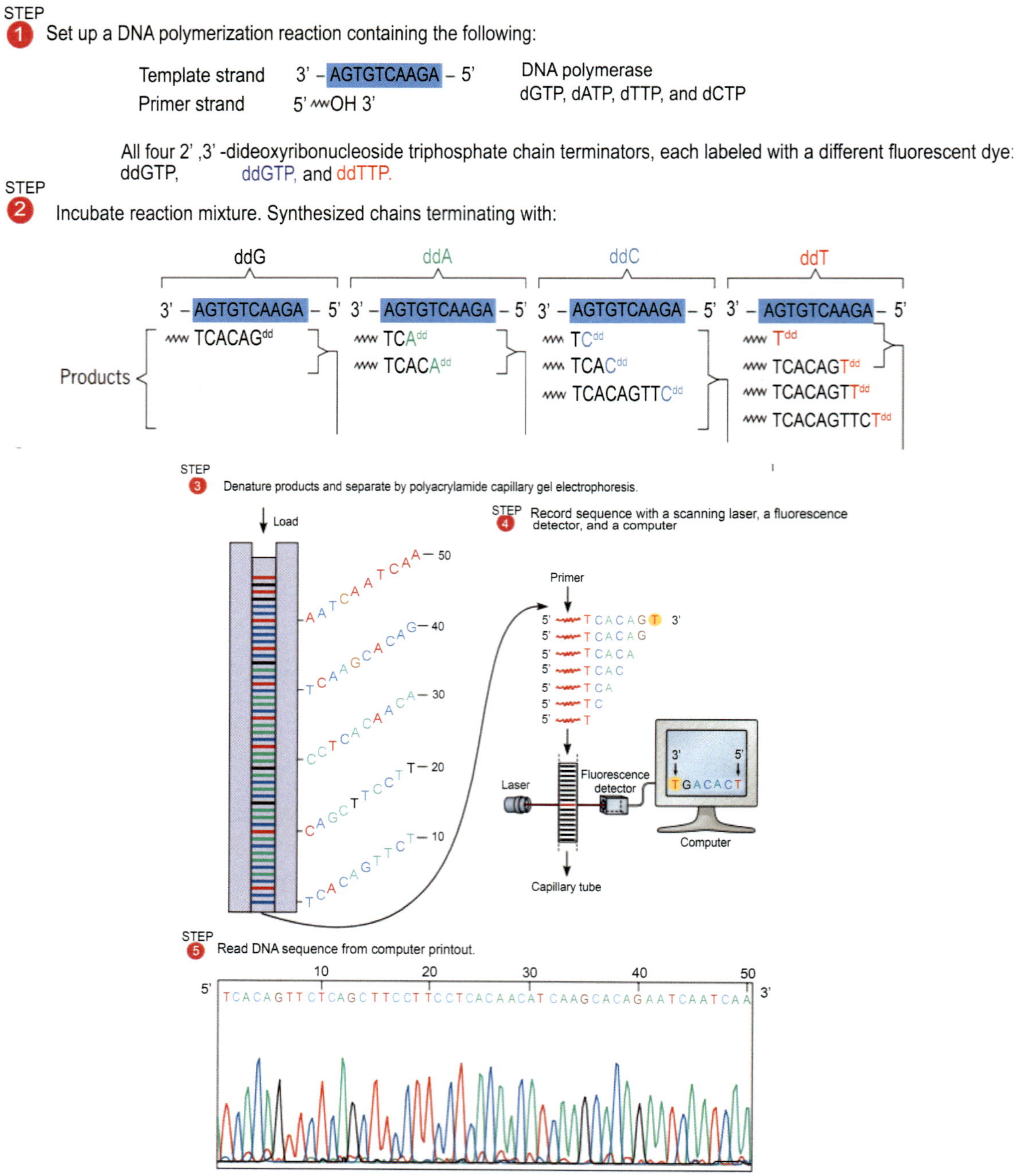

Figure 7.17. *Sequencing DNA by the 2,3-dideoxynucleoside triphosphate chain-termination procedure.*

Source: https://www.google.co.in/books/edition/Principles_of_Genetics/NBB0CgAAQBAJ?hl=en&gbpv=1&dq=genetics+book&printsec=frontcover.

In vitro DNA synthesis is performed in the presence of the four 2,3-dideoxy chain terminators: ddGTP, ddATP, ddCTP, and ddTTP, each labeled with a different fluorescent dye. The reaction mixture contains all the components required for DNA synthesis. The dideoxy terminator at the 3 end of each chain is determined by the fluorescence of the attached dye. In the example shown, ddG fluoresces dark blue (appears black), ddC fluoresces light blue, ddA fluoresces green, and ddT fluoresces red. Because the shortest chain migrates the greatest distance, the nucleotide sequence of the longest chain (shown reading 5 → 3 at the top of the computer printout) is obtained by reading the sequence starting with the first chain to pass the laser beam and continuing with each chain one nucleotide longer through to the longest chain.

According to Figure 7.17, all chains that end in ddG will fluoresce one color, all chains that end in ddA will fluoresce two colors, all chains that end in ddC will fluoresce three colors, and all chains that end in ddT will fluoresce four colors. The ratio of dXTP:ddXTP (where X can be any of the four bases) is maintained at roughly 100:1 in the reaction tube, meaning that the likelihood of the nascent chain terminating at a specific X is roughly 1/100. As a result, a population of fragments is produced that ends within a few 100 nucleotides of the original primer terminus at every possible (X) termination site. Following denaturation, which releases the DNA chains produced in the reaction from the template strands, the DNA chains are separated using polyacrylamide capillary gel electrophoresis. A scanning laser and a fluorescence detector are used to locate the DNA chains in the gel, and their positions are then recorded on a computer. The series of fluorescence peaks that are captured as each developing chain passes the laser beam is printed out by the computer. Every chain that passes through the gel after the shortest one does so with one nucleotide longer than the last. The color of the fluorescence is determined by the dideoxynucleotide at the end of each chain. As a result, reading the fluorescence peak sequence from the shortest chain to the longest chain will reveal the sequence of the longest newly synthesized DNA chain (Figure 7.17). To assess your knowledge of automated DNA sequencing devices that employ the Sanger technique, see Problem-Solving Skills: Determining the Nucleotide Sequences of Genetic Elements.

Remember

Because it depends on the detection of the pyrophosphate released when a nucleotide is added to the end of a primer strand, one such technique is known as pyrosequencing.

The Sanger chain terminator method is currently being replaced by new methods for sequencing DNA, and these new, so-called second generation, machines are capable of sequencing up to 25 billion nucleotide pairs every day. Numerous novel sequencing techniques make use of sequencing-by-synthesis protocols, in which DNA polymerase extends the primer strands of immobilized primer-template complexes by adding deoxyribonucleoside triphosphates one at a time. The sequence of nucleotide additions is then recorded using light signals captured by a charge-coupled device (CCD) sensor.

In a different process, primer strands attached to tiny beads in a water-oil mixture are extended, and the addition of fluorescently labeled nucleotides is recorded using a laser beam. We refer to this process as 454 sequencing. Reversible terminators are used in a different process known as Illumina sequencing (formerly Solexa sequencing) to identify individual nucleotides as they are added to developing DNA strands. This process is often referred to as massively parallel sequencing because it involves a large number of reactions happening at once in the sequencing machines that use it. These systems are all incredibly quick, and fresh sequencing techniques are always being created. The goal of sequencing the entire human genome for $1,000 has moved from science fiction to a realistic possibility, even though we are still a long way off.

SUMMARY

- Genetics is the study of the inheritance and variation of biological traits. We have previously noted that it is possible to conduct genetic research without directly studying DNA.
- DNA cloning is a molecular biology technique which is used for the creation of exact copies or clones of a particular gene or DNA. DNA cloning is the process of making multiple copies of a particular segment of DNA.
- A restriction endonuclease catalyzes the cleavage of a specific sequence of nucleotide pairs regardless of the source of the DNA. It will cleave phage DNA, *E. coli* DNA, corn DNA, human DNA, or any other DNA, as long as the DNA contains the nucleotide sequence that it recognizes.
- Genomic DNA libraries are usually prepared by isolating total DNA from an organism, digesting the DNA with a restriction endonuclease, and inserting the restriction fragments into an appropriate cloning vector.
- The development of recombinant DNA techniques has spawned many new approaches to the analysis of genes and gene products.
- Gel electrophoresis is a powerful tool for the separation of macromolecules with different sizes and charges. DNA molecules have an essentially constant charge per unit mass; thus, they separate in agarose and acrylamide gels almost entirely on the basis of size or conformation.
- Polyacrylamide gel electrophoresis is an important tool for the separation and characterization of proteins. Because many functional proteins are composed of two or more subunits, individual polypeptides are separated by electrophoresis in the presence of the detergent sodium dodecyl sulfate (SDS), which denatures the proteins.

MULTIPLE CHOICE QUESTIONS

1. Which of the following enzymes is used to amplify a specific DNA sequence in PCR?

a. DNA ligase

b. Restriction enzyme

c. DNA polymerase

d. RNA polymerase

2. What does gel electrophoresis separate DNA fragments based on?

a. Sequence

b. Size and charge

c. Shape

d. Solubility

3. Which technique is used to detect specific RNA sequences?

a. Southern blotting

b. Western blotting

c. Northern blotting

d. Eastern blotting

4. In which technique is a DNA probe used to identify specific DNA sequences?

a. PCR

b. Southern blotting

c. Western blotting

d. ELISA

5. What is the purpose of the "Sanger sequencing" method?

a. To separate proteins by size

b. To quantify RNA levels

c. To determine the nucleotide sequence of DNA

d. To visualize protein-protein interactions

Answers to Multiple Questions

1. (c); 2. (b); 3. (c); 4. (b); 5. (c)

REVIEW QUESTIONS

1. Discuss the production of recombinant DNA molecules *in vitro*
2. How does the use of different gel concentrations affect the resolution of gel electrophoresis?
3. How does the process of cloning involve restriction enzymes and DNA ligase?
4. What is the difference between quantitative PCR (qPCR) and traditional PCR?
5. What are the advantages and limitations of using CRISPR-Cas9 for gene editing?

REFERENCES

1. Bonner, J. (1965). *The Molecular Biology of Development*. Oxford: Oxford University Press.
2. Burian, R. M. (1986). On conceptual change in biology: The case of the gene. In D. J. Depew & B. H. Weber (Eds.), *Evolution at a Crossroads* (pp. 21–42). Cambridge, MA: MIT Press.
3. Burian, R. M. (1996). Underappreciated pathways toward molecular genetics as illustrated by Jean Brachet's cytochemical embryology. In S. Sarkar (Ed.), *The Philosophy and History of Molecular Biology: New Perspectives* (pp. 67–85). Dordrecht: Kluwer.

4. Collier, J. (1988). Supervenience and reduction in biological hierarchies. In M. Matthen & B. Linsky (Eds.), *Philosophy and Biology: Canadian Journal of Philosophy Supplementary* (Vol. 14, pp. 209–234).

5. Fan, J. B., Chikashige, Y., Smith, C. L., Niwa, O., Yanagida, M., & Cantor, C. R. (1989). Construction of a Not I restriction map of the fission yeast *Schizosaccharomyces pombe* genome. *Nucleic Acids Research*, *17*, 2801–2818.

6. Gelfand, M. S., & Koonin, E. V. (1997). Avoidance of palindromic words in bacterial and archaeal genomes: A close connection with restriction enzymes. *Nucleic Acids Research*, *25*, 2430–2439.

7. Hirai, H. (2020). Chromosome dynamics regulating genomic dispersion and alteration of nucleolus organizer regions (NORs). *Cells*, *9*(4), 1–14.

8. Pinto-Basto, J., Guimarães, B., Rantanen, E., Javaher, P., Nippert, I., Cassiman, J. J., Kääriäinen, H., Kristoffersson, U., Schmidtke, J., & Sequeiros, J. (2010). Scope of definitions of genetic testing: Evidence from a EuroGentest survey. *Journal of Community Genetics*, *1*(1), 29–35.

9. Primrose, S. B., & Twyman, R. M. (2006). *Principles of Genome Analysis & Genomics* (3rd ed.). Blackwell Publishing.

10. Smith, C. L., Econome, J. G., Schutt, A., Klco, S., & Cantor, C. R. (1987). A physical map of the *Escherichia coli* K12 genome. *Science*, *236*, 1448–1453.

11. Snustad, D. P., & Simmons, M. J. (2005). *Principles of Genetics* (6th ed.). John Wiley & Sons.

12. Wan, T. S., Hui, E. K., & Ng, M. H. (2017). Chromosome recognition. In *Methods in Molecular Biology* (Vol. 1541, pp. 67–74). Humana Press.

13. Wang, J., Liao, J., Zhang, J., Cheng, W. Y., Hakenberg, J., Ma, M., Webb, B. D., Ramasamudram-Chakravarthi, R., Karger, L., Mehta, L., Kornreich, R., Diaz, G. A., Li, S., Edelmann, L., & Chen, R. (2015). ClinLabGeneticist: A tool for clinical management of genetic variants from whole exome sequencing in clinical genetic laboratories. *Genome Medicine*, 7(1), 77.

CHAPTER

8 Genomics

LEARNING OBJECTIVES

After studying this chapter, you will be able to:

- Give an overview of genomics
- Focus on genome organization and sequencing techniques
- Explain the comparative genomics and functional genomics
- Discuss the genome editing techniques (e.g., CRISPR-Cas9)
- Learn about applications of genomics

INTRODUCTORY EXAMPLE

An Introduction to Genomic Medicine and Genetics in Health Care

Genetics plays a pivotal role in healthcare, influencing everything from disease risk assessment to personalized treatment strategies. At its core, genetics is the study of genes and their impact on traits, health, and disease. Understanding genetics in healthcare involves exploring how variations in genes can predispose individuals to certain conditions, how genetic testing can inform diagnosis and treatment decisions, and how genetic counseling can help individuals and families navigate genetic information.

What Is Genomic Medicine?

Genomic medicine is the study of our genes (DNA) and their interaction with our health. Genomics investigates how a person's biological information can be used to improve their clinical care and health outcomes (e.g., through effective diagnosis and personalized treatment.

While genetics looks at specific genes or groups of 'letters' along the DNA strand, genomics refers to the study of someone's entire genetic makeup. It is about how they relate and react with each other and is associated with conditions that have a broader range of triggers such as diabetes, heart disease, cancer, and asthma.

How Is Genomics Used in Medicine?

- Diagnosis — for example, where the cause of a range of symptoms cannot be pinpointed by any other means.
- Prenatal tests that take place during pregnancy — either to screen (just in case something is wrong with the baby) or where there is already a family history. It helps the parents to make informed choices and plans for the future.
- Where there is a family history of serious genetic disorders, it can tell prospective parents whether or not they are a carriers and if they can pass it on to their children. It can also tell someone if they are likely to develop the inherited condition later in life, even if they don't yet have any symptoms.
- To assess risk — someone's genetic makeup can show their susceptibility to suffering certain illnesses, like heart disease, stroke, and cancer. Perhaps they're likely to have high cholesterol levels or to suffer problems with their veins. Possessing this knowledge means they can manage the risk through medicines, medical intervention, or making positive lifestyle changes.
- The Ways in Which Genomic Medicine Is Making a Difference:

This greater understanding of the links between biology and disease brings benefits on several levels.

- Personal — each patient has medicines, treatment, and a health care plan tailored to them and their individual needs and risks. As an example, take the treatment of colorectal cancers. Some people with a particular gene mutation have better survival rates when treated with a non-steroidal anti-inflammatory, such as aspirin than those without this mutation.

- Doctors — access to genomic information helps with diagnosis, managing treatments, and spotting symptoms across a wider cohort of patients. There have been a few cases where cerebral palsy diagnosis has been re-evaluated in the light of genetic testing, revealing a new diagnosis and, as a result, a new, effective treatment plan.
- National level — developing strategies to care for rising trends and particular communities and programs like newborn screening in the U.S., which examines between 29 and 50 severe but treatable conditions.
- On a world-wide scale — projects like the Online Mendelian Inheritance in Man, an open-access database of all known human genetic conditions. This kind of approach means that the parents of children with rare syndromes are more likely to get the answers and the support they need.

What Can I Expect and What Is Genetic Counseling?

There are a number of types of service providers. In the U.K., for example, the National Health Service employs 90 consultant clinical geneticists at 25 centers. They're supported by hundreds of specifically trained staff. Referral is usually through a general practitioner (GP or family doctor) and is available to those who are worried about a serious genetic family condition or a family tendency towards developing cancer, or to parents of a child with learning difficulties and other developmental problems looking for an expert assessment.

In places where a public service isn't available, or for those who choose to seek private health care treatment, check to make sure that the clinic you're using has the necessary registration (for example, in the UK this is through the Care Quality Commission, also known as the CQC) and the lab is also correctly accredited.

Whatever the setting, the appointment might take some time and you may need to bring other members of your family with you. Your family and medical history will be mapped and explored, and it is likely you'll have a medical examination too. Finding out that there may be a life-changing or life-limiting condition in your future is a serious and, for some, traumatic experience. Alongside counseling, you may be offered tests (including blood tests)—with the option of having these done on the day or, if you need time to think about the possible implications, to come back at a later date.

Results can take weeks or even months to return (depending on the rarity of the genetic abnormality and how easy it is to find) but pre-natal test results will be returned much sooner.

Aftercare then depends on the results and the nature of what you're being tested for. Some people will be referred back to their family doctor along with full details, or they may go on to receive treatment at a specialist unit. Those who aren't showing symptoms will be given support and advice about lifestyle changes, in order to minimize their risk and advice about managing their potential condition in the future.

There are also a number of private companies that offer genetic testing by mail. It involves having a cheek swab or a blood sample taken at a local clinic. It is then sent off to the laboratory. The kinds of things tested for include genetic risk for diabetes and heart conditions, as well as ancestry information. Some companies deliver more of a service than others, with counselors or other health professionals on hand to help. Convenient (but not necessarily cheap), it must be remembered that this is genetic testing without the usual level of holistic support found in established clinics.

The Future:

The broad area known as genomic medicine is evolving — the study of genetic mutation pathways and their variations is particularly exciting. But what does this mean for people on a practical level? As discussed earlier, some hereditary diseases are difficult to diagnose simply because of the wide range of genes involved.

Scientists are working towards finding a chemical or genetic bottleneck for conditions like these. The ability to switch off a vital reaction along the pathway from genetic trigger to hay fever, dust allergy, or asthma, for example, would aid diagnosis and treatment, and possibly whether or not these traits would cause misery for the next generation.

UNIT INTRODUCTION

Biology has been completely transformed by genomics, the study of an organism's whole genome, which has given researchers previously unheard-of insights into the composition, evolution, and structure of genetic material. Fundamentally, genomics is the study of DNA sequences in their entirety, including all of the genes, regulatory components, and non-coding regions that make up an organism's genome. The Human Genome Project (HGP), which resulted in the first draft sequence of the human genome, was completed in 2003 and is considered one of the foundational achievements in genomics. The genomic era began with this enormous project, which allowed scientists to investigate the intricacies of human biology at the molecular level. Since then, developments in computational tools and sequencing technologies have made genomics a dynamic, quickly developing field.

- Applications for genomes are numerous and span many academic fields. Genomic research has transformed disease treatment, personalized medicine, and diagnostics in the medical field. Clinicians can predict medication responses, find genetic predispositions to diseases, and customize treatments for each patient by analyzing their individual genomes. Targeted therapies and precision medicine approaches are made possible by the understanding of the genetic basis of complex diseases, including neurological conditions, cancer, and cardiovascular disorders, which is made possible by genomics.

Genes are driving innovations in sustainable agriculture, livestock breeding, and crop improvement. Researchers can discover genes linked to desired qualities like yield, disease resistance, and nutritional content by sequencing the genomes of crop plants and livestock species. They can then utilize this knowledge to create improved varieties with increased resilience and productivity. Genomic analysis sheds light on the genetic diversity, adaptability, and evolutionary history of species in the fields of ecology and evolutionary biology. Through the use of comparative genomics, scientists can investigate the molecular mechanisms underlying evolutionary change as well as the genetic underpinnings of phenotypic variations between species. Additionally, the functioning of ecosystems, environmental adaptation, and the dynamics of microbial communities are clarified by genomic studies.

Gene engineering, drug discovery, and biomanufacturing are among the fields in which genomics finds application in biotechnology, bioinformatics, and synthetic biology. Through the use of genomics, researchers are opening up new areas of study in the fields of agriculture, environmental science, and biomedicine. These discoveries have significant effects on environmental sustainability, food security, and human health. The field of genomics has great potential to solve today's most important social issues and solve life's mysteries as it develops.

8.1. OVERVIEW OF GENOMICS

Learning Objectives

- Discuss the history of genomics;
- Explain the types of genomics;
- Focus on genome analysis.

Genomics is the study of an organism's entire genome, combining aspects of genetics. To sequence, assemble, and analyze the structure and function of genomes, genomics combines recombinant DNA, DNA sequencing techniques, and bioinformatics. It is distinct from "classical genetics" in that it takes into account all of an organism's genetic components as opposed to focusing on a single gene or gene product at a time. Additionally, the study of genomics focuses on interactions that occur within the genome between loci and alleles as well as other interactions like heterosis, pleiotropy, and epistasis (Figure 8.1). The field of genomics, which utilizes whole DNA sequences for entire organisms, was made possible by the combination of Fred Sanger's groundbreaking work and the more recent advances in next-generation sequencing technology.

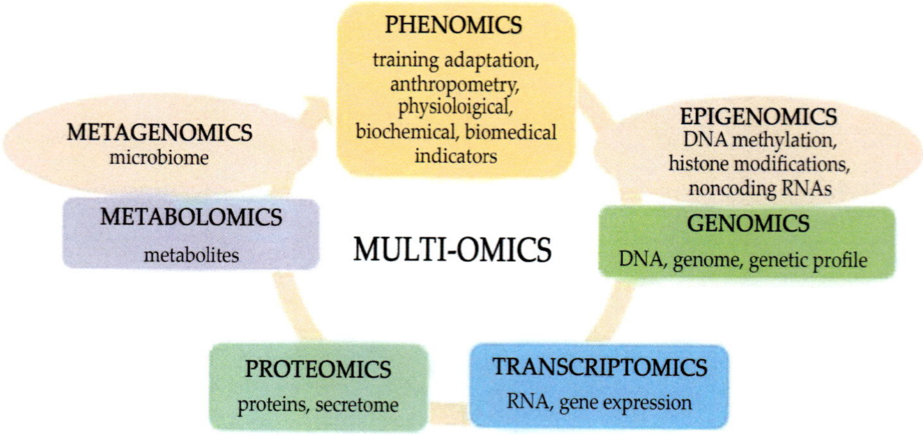

Figure 8.1. *Genomics studies the genomes of whole organisms and other intragenomic interactions.*

Source: Ginevičienė, V., Utkus, A., Pranckevičienė, E., Semenova, E. A., Hall, E. C. R., & Ahmetov, II, (2022). Perspectives in sports genomics. Biomedicines, 10(2), 298. https://doi.org/10.3390/biomedicines10020298.

8.1.1. History of Genomics

Even though DNA was first isolated in 1869, the first genomes were not sequenced for over a century, making genomics a relatively new field of study. Although there were a number of important turning points that shaped the field in the century before, genomics as we know it today really started in the 1970s.

8.1.1.1. Early History (1971–1910): Discovery of DNA

In a paper published in 1871, Friedrich Miescher discovered that the nucleus of cells contains nuclein and related proteins. This is what we now call DNA, and it serves as the foundation for the study of genomics. Theodor Boveri and Walter Sutton discovered in 1904 that chromosomes are inherited from both parents and occur in matched pairs. The Chromosome Theory of Heredity was the term given to this.

Albert Kossel was awarded the Nobel Prize in Physiology and Medicine in 1910 for the discovery of the five nucleotide bases, which are:

- Adenine (A);
- Cytosine (C);
- Guanine (G);
- Thymine (T); and
- Uracil (U).

Keyword

Heredity – It refers to the genetic heritage passed down by our biological parents.

8.1.1.2. 1950–1968: Development of Knowledge About DNA

In 1950, Erwin Chargaff advanced our knowledge of how the nucleotide bases adenosine, cytosine, guanine, and thymine pair together. He found that in DNA samples, the ratios of thymine to adenine or cytosine to guanine were always the same. He therefore concluded that adenosine and thymine form chromosome pairs, while cytosine and guanine form chromosome pairs.

Through a series of experiments, Alfred Hershey and Martha Chase proved in 1952 that genetic traits that are passable down through the generations are carried by DNA, not by proteins. In 1953, Francis Crick and James Watson discovered DNA's double helix structure.

The research team led by Marshall Nirenberg and Har Gobind Khorana in 1961 discovered what is now referred to as the "codon" of DNA. In DNA, this sequence of three letters stands for a particular amino acid. Specific proteins are created during translation, which is when amino acids come together. Together with Robert Holley, who sequenced the first transfer RNA (tRNA) molecule, they were awarded the 1968 Nobel Prize for Physiology and Medicine for their work deciphering the genetic code.

8.1.1.3. 1977–Present: Sequencing of DNA

Frederick Sanger created the phiX174 virus in 1977 using a DNA sequencing technique, which led to the first complete genome being sequenced and created new opportunities in the field of genomics. Along with Wally Gilbert and Paul Berg, he was awarded the Nobel Prize in 1980 for developing the techniques for DNA sequencing. James Gusella discovered the location of the gene linked to Huntington's disease in 1983. The polymerase chain reaction (PCR), a method for amplifying DNA, was created by Dr. Kary Mullis that same year. Alec Jeffreys created a technique in 1985 for profiling DNA by calculating the number of repetitive DNA sequences in particular genome regions.

Methods for examining developing embryos in the womb for genetic disorders were created in 1992. This allowed expectant parents to determine whether hemophilia or cystic fibrosis was likely to affect their unborn child. Further, *Haemophilus influenza* was the first bacterium whose genome was fully sequenced in 1995. The genome of the yeast *Saccharomyces cerevisiae* was finished a year later. This was also the birth of Dolly the Sheep, the first cloned animal. The genome of the nematode worm *Caenorhabditis elegans* was published in 1998 by John Sulston and Bob Waterston.

The goal of the 1990-launched Human Genome Project was to sequence every one of the 3 billion letters that make up the human genome. The first chromosome to be sequenced as part of this project in 1999 was chromosome 22. When the project was finished in 2003, it was determined that humans have between 20,000 and 25,000 genes. In 2002, the first mammalian full genome sequence was obtained by the International Mouse Genome Sequencing Consortium. Although the mouse's genome is smaller than the human genome, it is nevertheless very similar. To compile a list of common variations found in the human genome, the International HapMap Project was also started this year. A breakthrough in DNA sequencing technology in 2007 resulted in a 70-fold increase in the amount of DNA sequenced in a single year. As a result, the 1000 Genes Project was started in 2008 with the goal of sequencing the genomes of 2,500 members of a sizable population group. 2010 saw the publication of this project's findings.

8.1.2. Types of Genomics

Numerous new branches have emerged, along with some intriguing findings. As a result, the field of genomics has subtly expanded. Nonetheless, structural genomics, functional genomics, and comparative genomics are the three main categories into which genomics falls.

8.1.2.1. Structural Genomics

Structural genomics is the study and application of genome sequencing. DNA sequencing, sequence assembly, sequence management, and sequence organization are all covered by structural genomics. Rather than focusing on a single protein, structural genomics aims to ascertain the structure of every protein that the genome encodes. In essence, it is the initial phase of the genome analysis, i.e., creation of high-resolution genetic or sequence maps of the organism. An organism's entire DNA sequence serves as its most comprehensive physical map. The idea of structural genomics has entered a transitional phase as a result of the quick development of DNA technology and the conclusion of multiple genome sequencing projects in recent years. As proteins differ among individuals and thus do genome sequences, structural genomics now encompasses the systematic determination of the three-dimensional structure of proteins found in living cells.

8.1.2.2. Functional Genomics

The foundation of functional genomics is the study and comprehension of gene function. Rebuilding genome sequences can help determine the function of individual genes based on structural genomics. It provides a general understanding of each gene's function and how it manifests in an organism. DNA chips, 2D gels, bioinformatics sequences, and other tools are among the various resources helpful for structural genomics. This data supports the design of experiments aimed at determining the roles played by a particular genome. The field of biological research has expanded thanks to the functional genomics approach. This approach relies on a methodical examination of a single gene or protein in relation to all genes. Thus, functional genomics is characterized by large-scale experimental methodologies. Therefore, novel information about the genome is provided by functional genomics. This facilitates our understanding of genes, protein functions, and interactions between proteins. The advancements in proteomics and microarray technology have made it possible to investigate the real-time expression of every gene in a cell or tissue under various environmental conditions, such as pH, temperature, etc.

8.1.2.3. Comparative Genomics

The noteworthy discovery was that complete genome sequences of cellular organisms became available. A third of the genes encoded on each genome were discovered to have no known or predictable function, e.g., about 40% of the genes in *E. coli* K12 are unknown in their function. Microbial proteins have a fairly uniform degree of evolutionary conservation, with roughly 70% of the gene products from each sequenced genome having homologs in different genomes. In contrast to paralogs, orthologs (the same gene in different species that connect) have functional annotations of proteins that can be transferred to better study organisms and help predict the function of these genes, from less-studied organisms that are related through duplication within their genome. For a deeper comprehension of genomes, comparative genomics becomes an effective method. Analysis of protein sets from fully sequenced genomes is one of the many facets of comparative genomics. For comparative genomics, general-purpose and organism-specific databases are utilized.

Did you know?

The field of genomics has rapidly advanced since the completion of the Human Genome Project in 2003. Initially, sequencing the entire human genome took over a decade and cost billions of dollars. Today, advancements in technology have made it possible to sequence a human genome in just a few days at a fraction of the cost. This has revolutionized the study of genetics and genomics, leading to breakthroughs in personalized medicine, disease research, and understanding of human evolution.

8.1.3. Genome Analysis

Genome projects consist of three steps after an organism has been chosen: sequencing the DNA, assembling the sequence to create a representation of the original chromosome, and annotating and analyzing that representation.

8.1.3.1. Sequencing

Sequencing was carried out in centralized facilities known as sequencing centers. These facilities ranged in size from large independent institutions like the Joint Genome Institute, which sequences dozens of terabases annually, to local molecular biology core facilities. Research laboratories in these centers are equipped with the expensive equipment and technical support required for the job. But as sequencing technology advances, a new class of efficient benchtop sequencers with quick turnaround times has become affordable for the typical academic lab. High-throughput (or next-generation) sequencing and shotgun sequencing are the two main categories into which genome sequencing techniques generally fall.

8.1.3.1.1. Shotgun Sequencing

A sequencing technique called shotgun sequencing is intended for the analysis of DNA sequences longer than 1000 base pairs, including whole chromosomes. It gets its name from an analogy with a shotgun's rapidly expanding, almost random firing pattern. Since gel electrophoresis sequencing can only be used for fairly short sequences (100 to 1000 base pairs), longer DNA sequences must be broken into random small segments which are then sequenced to obtain reads. Several rounds of this fragmentation and sequencing yield multiple overlapping reads for the target DNA. The overlapping ends of various reads are then used by computer programs to put them together into a continuous sequence. The average number of reads by which a genome is over-sampled is known as coverage. Shotgun sequencing is a random sampling technique that necessitates over-sampling to guarantee a specific nucleotide is represented in the reconstructed sequence.

The classical chain-termination method, also known as the "Sanger method," was the technology behind shotgun sequencing for a large portion of its history. It is based on DNA polymerase's selective incorporation of chain-terminating dinucleotides during *in vitro* DNA replication. High-throughput sequencing techniques have recently replaced shotgun sequencing, particularly for large-scale, automated genome analyzes.

For smaller-scale projects and to obtain particularly long contiguous DNA sequence reads (>500 nucleotides), the Sanger method is still widely used, though. A single-stranded DNA template, a DNA primer, a DNA polymerase, normal deoxynucleoside triphosphates (dNTPs), and modified nucleotides (dideoxyNTPs) that stop DNA strand elongation are needed for chain-termination techniques. Since these chain-terminating nucleotides lack the 3⊠-OH group necessary for the creation of a phosphodiester bond between two nucleotides, when a ddNTP is incorporated, DNA polymerase stops extending DNA. ddNTPs can be labeled with fluorescence or radioactivity to be detected by DNA sequencers. These devices can typically sequence up to 96 DNA samples in a single batch (run) in as little as 48 hours.

8.1.3.1.2. *High-Throughput Sequencing*

High-throughput sequencing technologies, which parallelize the sequencing process and produce thousands or millions of sequences at once, are a result of the increased demand for inexpensive sequencing. The goal of high-throughput sequencing is to reduce DNA sequencing costs above and beyond what can be achieved with conventional dye-terminator techniques. As many as 500,000 sequencing-by-synthesis operations can be carried out in parallel during ultra-high-throughput sequencing.

Pascal Mayer and Laurent Farinelli created the Illumina dye sequencing technique in 1996 at the Geneva Biomedical Research Institute. It is based on reversible dye-terminators. In this procedure, local clonal colonies—originally dubbed "DNA colonies"—are created by first attaching primers and DNA molecules to a slide, then amplifying the results using polymerase. Reversible terminator bases (RT-bases) of four different types are added, and non-incorporated nucleotides are removed, in order to ascertain the sequence. In contrast to pyrosequencing, sequential images captured from a single camera can capture very large arrays of DNA colonies because the DNA chains are extended one nucleotide at a time and image acquisition can be done at a later time. Separating the enzymatic reaction from the image capture process enables optimal throughput and potentially infinite sequencing capacity; in an ideal setup, the camera's A/D conversion rate is the only factor determining the instrument's maximum throughput. The next cycle is made possible by chemically removing the dye and the terminal 3⊠ blocker from the DNA after the camera captures pictures of the fluorescently labeled nucleotides.

Keyword

Pyroseque-ncing – It is a method of DNA sequencing that differs from Sanger sequencing, in that it relies on the detection of pyrophosphate release and the generation of light on nucleotide incorporation, rather than chain termination with dideoxynucleotides.

Standard DNA replication chemistry serves as the foundation for an alternate strategy called ion semiconductor sequencing. With this technology, the release of a hydrogen ion after each base incorporation is measured. When a single nucleotide is added to a microwell containing template DNA, it is incorporated and a hydrogen ion is released if the nucleotide is complementary to the template strand. An ISFET ion sensor is triggered by this release. A homopolymer in the template sequence will allow for the incorporation of multiple nucleotides in a single flood cycle, resulting in a proportionally higher detected electrical signal.

8.1.3.2. Sequence Assembly

The process of aligning and combining segments of a much longer DNA sequence to recreate the original sequence is known as sequence assembly. This is necessary because, depending on the technology being used, current DNA sequencing methods can only read short segments of genomes, ranging from 20 to 1000 bases, rather than entire genomes as a continuous sequence. Sequencing reads longer than 10 kilobases are regularly produced by third-generation sequencing technologies like Oxford Nanopore and PacBio, but their error rate is high—roughly 15%. The short segments, known as reads, are usually the product of either gene transcripts (ESTs) or genomic DNA shotgun sequencing.

8.1.3.2.1. Assembly Approaches

In general, there are two methods for assembly: comparative assembly, which uses the current sequence of a closely related organism as a reference during assembly, and de novo assembly, which is used for genomes that are not similar to any that have previously been sequenced. De novo assembly is computationally challenging (NP-hard) in comparison to comparative assembly, which makes it less advantageous for short-read NGS technologies. There are two main strategies for assembly in the de novo assembly paradigm: overlap-layout-consensus (OLC) strategies and Eulerian path strategies. In the end, OLC strategies attempt to solve the NP-hard problem of finding a Hamiltonian path through an overlap graph. Since Eulerian path strategies search for an Eulerian path through a de Bruijn graph, they are more tractable in terms of computation.

8.1.3.2.2. Annotation

Without further analysis, the DNA sequence assembly by itself is not very useful. Annotating genomes, which involves three primary steps, allows biological information to be added to sequences.

i. identifying portions of the genome that do not code for proteins;
ii. identifying elements on the genome, a process called gene prediction; and
iii. attaching biological information to these elements.

Unlike manual annotation (a.k.a. annotation by hand), automatic annotation tools attempt to carry out these steps in silico. k. a. curation), which entails both potential experimental verification and human expertise. These methods should ideally work in tandem and enhance one another within the same annotation pipeline. Traditionally, BLAST is used to identify similarities before homologs are used to annotate genomes at the basic level of annotation. The annotation platform has recently received more information. Manual annotators can deconvolute differences between genes that have the same annotation thanks to the additional information. Certain databases provide genome annotations using their Subsystems approach by utilizing experimental data, genome context information, similarity scores, and resource integrations. Additional databases (e.g., Ensembl) use a variety of software tools in their automated genome annotation pipeline in addition to carefully selected data sources. The identification of genomic elements, mainly ORFs and their location, or gene structure makes up structural annotation. Functional annotation is the process of affixing biological data to genomic components.

Did you get it?

1. **Describe the Structural Genomics**
2. **Define the Functional Genomics**

8.2. GENOME ORGANIZATION AND SEQUENCING TECHNIQUES

The distribution of genes, regulatory elements, repetitive sequences, structural features, and other DNA sequences within an organism's genome is referred to as its genome organization. Deciphering genetic information, researching gene regulation, and examining the connections between genotype and phenotype all depend on an understanding of the organization of the genome. The size, structure, and complexity of genomes can differ greatly amongst different organisms. Prokaryotic genomes, which are found in bacteria and archaea, are usually smaller and simpler, with the majority of the genetic material in the organism being contained in a single circular chromosome. On the other hand, eukaryotic genomes, which are found in protists, fungi, animals, and plants, are frequently larger and more complex, consisting of several linear chromosomes that are housed inside the nucleus.

The entire genome of an organism can be deciphered using genome sequencing techniques, which are strong tools. The ability to quickly and affordably sequence entire genomes has transformed genomics research during the last few decades thanks to advancements in sequencing technologies.

There are several methods for genome sequencing, each with its own strengths and limitations:

1. **Sanger Sequencing:** This which Frederick Sanger invented in the 1970s, was the first widely used technique for sequencing DNA. It entails using DNA polymerase and chain-terminating dideoxynucleotides (ddNTPs) to synthesize DNA fragments of different lengths, then sorting the fragments according to size using gel electrophoresis. Sanger sequencing is very accurate, but its applicability to large-scale sequencing projects is limited by its cost and slowness.
2. **Next-Generation Sequencing (NGS):** High-throughput sequencing, or NGS technologies, allow millions of DNA fragments to be sequenced simultaneously in a massively parallel fashion. When compared to Sanger sequencing, these methods—Illumina, Ion Torrent, and Pacific Biosciences sequencing—offer quicker turnaround times and lower costs. Large-scale genome sequencing initiatives like the Human Genome Project and the 1000 Genomes Project have benefited greatly from NGS, which has transformed genomics research.

3. **Third-Generation Sequencing:** Large DNA fragments can be sequenced in real-time without the need for PCR amplification using third-generation sequencing technologies like single-molecule real-time (SMRT) and nanopore sequencing, which provide advantages over next-generation sequencing (NGS). These systems are able to produce long-read sequencing data, which are especially helpful for identifying structural variants, resolving complex genomic regions, and researching epigenetic changes.

Apart from genome sequencing, additional methods like chromosome conformation capture (3C) and its variations (e.g., Hi-C) are employed to investigate how genomes are arranged in three dimensions inside the nucleus. These techniques shed light on the spatial arrangement of genes and regulatory elements, as well as the architecture of the genome and chromatin structure. All things considered, genome organization and sequencing methods are vital resources for researching the composition, dynamics, and evolution of genomes as well as for use in biotechnology, agriculture, medicine, and conservation. It is anticipated that future developments in computational tools and sequencing technologies will further expedite genomic research and pave the way for novel discoveries.

8.3. COMPARATIVE GENOMICS AND FUNCTIONAL GENOMICS

Two complementary methods used in genomics research to examine the composition, operation, and evolution of genomes are comparative genomics and functional genomics.

8.3.1. Comparative Genomics

In comparative genomics, genomic sequences and structures from various species are analyzed to find commonalities, divergences, and evolutionary connections. Researchers can learn more about gene content, evolutionary processes, and genome organization by comparing genomes. Understanding the genetic basis of phenotypic traits and adaptations, identifying conserved genes and regulatory elements, and identifying evolutionary changes such as gene duplications and losses are some of the main goals of comparative genomics.

Utilizing bioinformatics tools and computational algorithms, comparative genomics aims to reconstruct evolutionary relationships, predict gene functions, and align and compare DNA sequences. Comparative analyzes encompass a wide range of tasks, from comparing closely related species pairwise to comparing entire genomes across different taxa. Numerous uses for comparative genomics exist, such as the identification of potential genes responsible for human diseases, the investigation of genome evolution and adaptation in model organisms, and the identification of novel genes and pathways with biotechnological or biomedical significance.

8.3.2. Functional Genomics

Deciphering the biological functions of genes and non-coding sequences within genomes, as well as their roles in development, disease, and cellular processes, is the goal of functional genomics. In order to comprehend how genes function in living things, functional genomics investigates gene expression, regulation, and interaction networks, in contrast to comparative genomics, which focuses on sequence similarities and evolutionary relationships.

Proteomics, which studies protein expression and interactions, transcriptomics, which studies gene expression patterns using tools like microarrays and RNA sequencing, and epigenomics, which studies epigenetic modifications like DNA methylation and histone modifications, are some of the experimental techniques used in functional genomics. Functioning genomics offers a methodical way to characterize protein interactions, regulatory networks, and gene expression profiles, which helps to understand the molecular mechanisms behind biological processes and diseases.

Keyword

Transcriptomics – It is the complete study of the transcriptome that includes all RNA molecules found within an organism.

A thorough understanding of genome structure, function, and evolution can be obtained by combining comparative genomics and functional genomics techniques. Functional genomics experiments can validate predictions from comparative analyzes and clarify the biological significance of genomic features, while comparative analyzes can identify candidate genes and regulatory elements for functional studies. When combined, these methods advance our knowledge of genome biology and its relevance to agriculture, environmental preservation, and public health.

8.4. GENOME EDITING TECHNIQUES (E.G., CRISPR-CAS9)

Because genome editing techniques like CRISPR-Cas9 allow for precise and targeted modifications to an organism's DNA sequence, they have revolutionized the fields of molecular biology and biotechnology. These methods can be applied in many different fields, such as bioproduction, gene therapy, agriculture, and basic research. The genome editing tool CRISPR-Cas9, also known as CRISPR-associated protein 9 and Clustered Regularly Interspaced Short Palindromic Repeats is derived from the bacterial immune system. It is composed of two primary parts:

1. **CRISPR RNA (crRNA):** This is a short RNA molecule that contains a sequence complementary to a target DNA sequence. The crRNA guides the Cas9 protein to the specific location in the genome that is to be edited.
2. **Cas9 Protein:** Cas9 is an enzyme that acts as a molecular scissors, capable of cutting DNA at precise locations specified by the crRNA. Once bound to the target DNA sequence, Cas9 creates a double-strand break (DSB) in the DNA.

The CRISPR-Cas9 system can be used to introduce various types of modifications to the genome:

1. **Gene Knockout:** By introducing DSBs in specific genes, CRISPR-Cas9 can disrupt gene function, resulting in a loss-of-function phenotype. This is commonly used to study gene function in research organisms or to develop disease models.
2. **Gene Editing:** In addition to causing DSBs, CRISPR-Cas9 can be used in conjunction with donor DNA templates to introduce precise changes to the genome, such as insertions, deletions, or substitutions. This enables the correction of disease-causing mutations or the introduction of desired genetic modifications.
3. **Gene Activation or Repression:** CRISPR-Cas9 can also be used to modulate gene expression without altering the underlying DNA sequence. By fusing Cas9 to transcriptional activators or repressors, researchers can selectively activate or silence target genes.

Compared to earlier genome editing methods, CRISPR-Cas9 offers a number of benefits, including simplicity, efficiency, and versatility. In many research labs, it has quickly emerged as the preferred method for genome editing, opening up a plethora of applications in biotechnology, agriculture, and biomedicine. But there are also moral and security issues with CRISPR-Cas9 use, especially when it comes to side effects, unexpected outcomes, and misuse potential. Research is still being done to address the ethical and legal issues surrounding the use of CRISPR-based genome editing techniques, as well as to increase the accuracy and safety of these techniques.

8.5. APPLICATIONS OF GENOMICS

Learning Objectives

- Focus on predicting disease risk at the individual level.
- Discuss the pharmacogenomics and toxicogenomics.
- Learning about microbial genomics: metagenomics.
- Explain the microbial genomics: creation of new biofuels

These days, genomics is applied to many different fields, including mitochondrial genomics, pharmacogenomics, and metagenomics. The most well-known use of genomics is in the study and discovery of medical treatments.

8.5.1. Predicting Disease Risk at the Individual Level

Genome analysis is used to screen currently healthy individuals at the individual level in order to predict the risk of disease. Before a disease manifests, medication and lifestyle modifications may be advised as an intervention. This method works best, though, when the issue is caused by a single gene defect. In developed nations, only about 5% of diseases are caused by such defects. The majority of common diseases, including heart disease, are polygenic, or multi-factored, meaning that they involve both environmental factors like diet and phenotypic characteristics involving two or more genes. To determine Quake's percentage of risk for 55 distinct medical conditions, a risk assessment was carried out. It was discovered that he had an uncommon genetic mutation that put him at risk for an unexpected heart attack. Additionally, it was estimated that he had a 1 in 4.4 chance of getting Alzheimer's disease and a 23% chance of getting prostate cancer. The scientists examined the genomic data using a number of publications and databases. The ethical concerns surrounding population-level genomic analysis persist, despite the increasing affordability of genomic sequencing and the improvement of analytical tools (Figure 8.2).

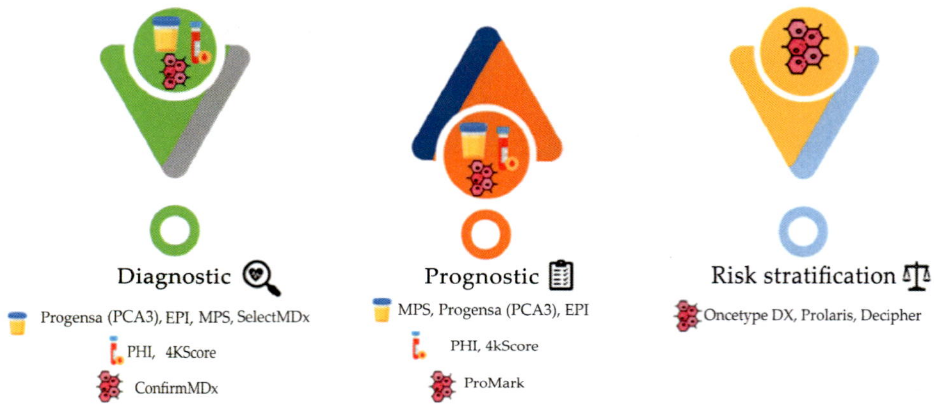

Figure 8.2. *PCA3 is a gene that is expressed in prostate epithelial cells and overexpressed in cancerous cells. A high concentration of PCA3 in urine is indicative of prostate cancer. The PCA3 test is considered to be a better indicator of cancer than the more well-known PSA test, which measures the level of PSA (prostate-specific antigen) in the blood.*

Source: Berenguer, C. V., Pereira, F., Pereira, J. A. M., & Câmara, J. S., (2022). Volatilomics: An emerging and promising avenue for the detection of potential prostate cancer biomarkers. Cancers, 14(16), 3982. https://doi.org/10.3390/cancers14163982.

Keyword

Genomic Sequencing – It provides data on the genetic changes in the virus that might impact vaccine efficacy.

8.5.2. Pharmacogenomics and Toxicogenomics

Pharmacogenomics, also referred to as toxicogenomics, is the study of drug safety and efficacy using data derived from a person's genetic sequence. Prior to starting studies on humans, the genomic responses to drugs can be investigated in the lab using live cells or experimental animals, such as mice or rats. Analyzing variations in gene expression may reveal details about the transcription profile while the medication is present, which may be utilized as a preliminary warning sign for potentially harmful side effects. For instance, disruption of genes involved in controlled cell death and proliferation may result in the development of cancerous cells. Finding new genes implicated in drug toxicity can also be aided by genome-wide research projects. Based on the genotype of each patient, personal genome sequence data can be used to prescribe drugs that will be most beneficial and least harmful. Even though the gene signatures might not be entirely correct, they can still be examined further before pathological symptoms develop.

8.5.3. Microbial Genomics: Metagenomics

Microorganisms are best studied under pure culture conditions, which entail isolating a single type of cell and culturing it in a laboratory, according to the traditional teaching of microbiology. Microorganisms can undergo multiple generations within a few hours, allowing their gene expression profiles to rapidly adjust to the novel laboratory conditions. Furthermore, almost all species of bacteria are resistant to being cultured in isolation. The majority of microorganisms reside in microbial communities called biofilms rather than as solitary organisms. Pure culture is not always the most effective method for studying microorganisms for the reasons listed above. The study of the combined genomes of several species that coexist and develop in a particular environmental niche is known as metagenomics. Metagenomics is a useful tool for analyzing the impact of pollutants on the environment and for quickly identifying new species (Figure 8.3).

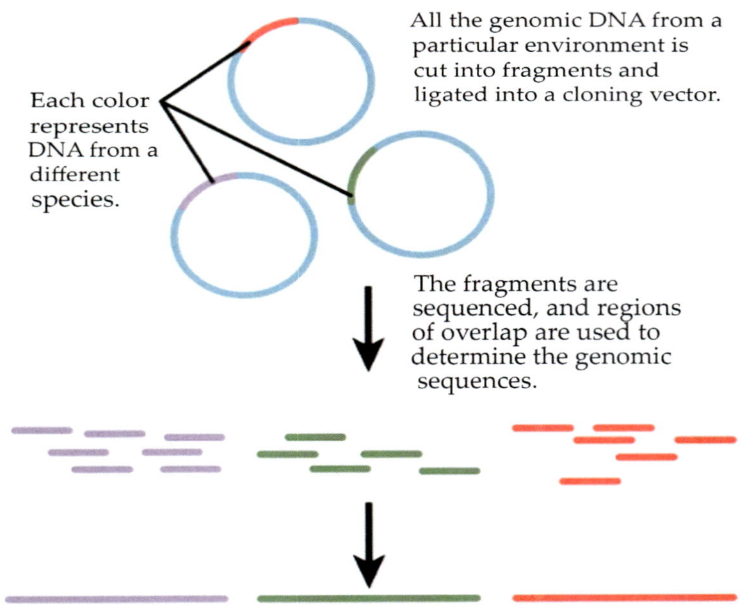

Figure 8.3. *Metagenomics involves isolating DNA from multiple species within an environmental niche.*

Source: https://upload.wikimedia.org/wikipedia/commons/1/1c/Figure_17_04_02.jpg.

8.5.4. Microbial Genomics: Creation of New Biofuels

Improved methods of extracting biofuels from algae and cyanobacteria are being developed with the aid of knowledge about the genomics of microorganisms. These days, wood, coal, oil, and other plant-based products like ethanol are the main sources of fuel. Even though plants are renewable resources, more alternative renewable energy sources must be discovered in order to meet the energy needs of our growing population. One of the greatest untapped sources of genes for encoding novel enzymes and synthesizing novel organic compounds is the microbial realm. Products like research-useful enzymes, antibiotics, and other anti-microbial mechanisms are produced by microorganisms. Microbial genomics is contributing to the development of advanced environmental cleanup methods, better vaccines, new disease treatments, and diagnostic tools.

8.5.4.1. Mitochondrial Genomics

Organelles found inside cells, mitochondria have their own DNA. Since mitochondrial DNA mutates quickly, it is frequently used to examine evolutionary relationships. The fact that the majority of multicellular organisms receive their mitochondrial DNA from their mothers during fertilization is another intriguing aspect of researching the mitochondrial genome. This is why tracing ancestry using mitochondrial genomics is common. Genetic markers have been used in forensic analysis, and information and hints gleaned from DNA samples recovered at crime scenes have been used as evidence in court cases. In this field, genomic analysis has also gained utility. The first publication utilizing genomics in forensics dates back to 2001. Academic research centers and the FBI worked together to try to solve the enigmatic anthrax cases that were sent via the US Postal Service. Researchers discovered that a particular strain of anthrax was utilized in each mailing by using microbial genomics.

8.5.4.2. Genomics in Agriculture

To a certain extent, genomics can lessen the trials and errors associated with scientific research, which could enhance the caliber and volume of agricultural yields. Crop breeding can be improved by linking traits to genes or gene signatures in order to produce hybrids with the most desired traits. By identifying desired traits from genomic data, scientists can subsequently transfer those traits to an organism of choice. Researchers are finding new applications of genomics to enhance both the quantity and quality of agricultural output. For instance, scientists could modify a crop that is sensitive to drought to become more tolerant of the dry season, or they could use desirable traits to improve an already-existing product.

Did you get it?

1. **Focus on mitochondrial genomics.**
2. **Discus the genomics in agriculture.**

ROLE MODEL

JAMES WATSON

James Dewey Watson is an American geneticist and biophysicist. He is noted for his decisive work in the discovery of the molecular structure of DNA, the hereditary material associated with the transmission of genetic information. He shared the Nobel Prize for Physiology or Medicine with Francis Crick and Maurice Wilkins in 1962.

Early Life and Education: James Watson was born on 6 April in 1928 in Chicago, Illinois. His father was a tax collector of Scottish ancestry. An only son, as a child he loved bird watching with his father.

He attended the University of Chicago, earning a degree in zoology in 1947. Watson was awarded a scholarship to Indiana University where he received his doctorate in 1950 for his work on the effects of X-ray radiation on replication of bacteriophage viruses.

Career Path: Watson then moved to Copenhagen, continuing with his virus research. Watson became interested in the deoxyribonucleic acid (DNA) molecule and joined a research team at the Cavendish Laboratory of the University of Cambridge with Francis Crick. Here, he made his famous discovery identifying the double helix structure of DNA.

He was appointed professor of biology at Harvard University in 1955, a position he held until 1976. His research at Harvard concentrated on RNA and its role in the transfer of genetic information.

He also became the director of Cold Spring Harbor Laboratory in New York in 1968, a position that he held for 35 years. Watson is credited with turning the laboratory into a world center for research in molecular biology.

Watson focused his research at the laboratory on exploring molecular biology and genetics in order to advance the understanding and ability to diagnose and treat cancers, neurological diseases, and other causes of human suffering.

In 1990, Watson was appointed as the Head of the Human Genome Project, a position he held for two years.

He retired in 2007 from the Cold Spring Harbor Laboratory, the same year as his fully sequenced genome was published online.

Contributions and Achievements: James Watson gained worldwide fame and prominence as the joint author of the four scientific papers between 1953 and 1954 (which he co-wrote with fellow scientist Francis Crick) that laid down the double helical structure of deoxyribonucleic acid (DNA), a megamolecule that is the fundamental substance in the process of genetic replication. This discovery won Watson and Crick (with Maurice Wilkins) the Nobel Prize in Physiology or Medicine in 1962.

During the 1960s, Watson became one of the most celebrated science writers, as he published his textbook "Molecular Biology of the Gene" in 1965 and his best-selling autobiographical book "The Double Helix" in 1968. Watson became the undisputed leading voice in the whole of American science. He epitomized the scientific creativity in 20th-century science, giving rise to molecular biology and its two applied offsets; biotechnology and the "Human Genome Project."

Watson married Elizabeth Lewis in 1968 and they had two sons, Rufus and Duncan.

His autobiography, a candid and entertaining memoir, "Avoid Boring People: Lessons from a Life in Science," published in 2007, is also filled with practical advice for those starting out their academic careers.

SUMMARY

- Gene engineering, drug discovery, and biomanufacturing are among the fields in which genomics finds application in biotechnology, bioinformatics, and synthetic biology.
- Structural genomics is the study and application of genome sequencing. DNA sequencing, sequence assembly, sequence management, and sequence organization are all covered by structural genomics.
- The foundation of functional genomics is the study and comprehension of gene function. Rebuilding genome sequences can help determine the function of individual genes based on structural genomics.
- Pharmacogenomics, also referred to as toxicogenomics, is the study of drug safety and efficacy using data derived from a person's genetic sequence. Before starting studies on humans, the genomic responses to drugs can be investigated in the lab using live cells or experimental animals, such as mice or rats.
- Microorganisms are best studied under pure culture conditions, which entail isolating a single type of cell and culturing it in a laboratory, according to the traditional teaching of microbiology.
- A thorough understanding of genome structure, function, and evolution can be obtained by combining comparative genomics and functional genomics techniques.

MULTIPLE CHOICE QUESTIONS

1. **Which of the following best describes genomics?**
   a. Study of individual genes
   b. Study of interactions between genes and the environment
   c. Study of the entire set of genes in an organism
   d. Study of genetic mutations
2. **What is the primary focus of comparative genomics?**
   a. Identifying genetic mutations in individuals
   b. Comparing gene expression patterns between different organisms
   c. Analyzing similarities and differences in the DNA sequences of different species
   d. Studying the structure and function of individual genes
3. **Which of the following techniques is commonly used for whole-genome sequencing?**
   a. Polymerase chain reaction (PCR)
   b. Sanger sequencing
   c. Next-generation sequencing (NGS)
   d. Gel electrophoresis
4. **What is the role of bioinformatics in genomics?**
   a. Studying the physical characteristics of DNA

b. Analyzing large-scale biological data using computational tools
c. Identifying mutations in specific genes
d. Creating genetic maps of organisms

5. Which of the following accurately describes a gene expression study using genomics techniques?

a. Identifying single nucleotide polymorphisms (SNPs) in the genome
b. Determining the three-dimensional structure of a protein
c. Analyzing changes in mRNA levels in response to different conditions
d. Sequencing the entire genome of an organism

REVIEW QUESTIONS

1. Define genomics and explain its significance in modern biology.
2. Describe the difference between genomics and genetics. How do they complement each other in studying inheritance and variation?
3. Explain the concept of a genome. What types of information can be found within a genome?
4. What are the key goals of genome sequencing projects, such as the Human Genome Project?
5. Discuss the importance of comparative genomics in understanding evolutionary relationships between species.

Answers to Multiple Questions

1. (c); 2. (c); 3. (c); 4. (b); 5. (c)

REFERENCES

1. De Luca, V., Viggiano, E., Dhoot, R., Kennedy, J. L., & Wong, A. H. (2009). Methylation and QTDT analysis of the 5-HT2A receptor 102C allele: Analysis of suicidality in major psychosis. *Journal of Psychiatric Research*, *43*, 532–537.
2. Dewey, F. E., Grove, M. E., Pan, C., Goldstein, B. A., Bernstein, J. A., Chaib, H., et al. (2014). Clinical interpretation and implications of whole-genome sequencing. *JAMA*, *311*(10), 1035–1045.
3. Grozeva, D., Kirov, G., Ivanov, D., Jones, I. R., Jones, L., Green, E. K., St Clair, D. M., et al. (2010). Welcome Trust Case Control Consortium: Rare copy number variants: A point of rarity in genetic risk for bipolar disorder and schizophrenia. *Archives of General Psychiatry*, *67*, 318–327.
4. Hudson, K. L. (2011). Genomics, health care, and society. *The New England Journal of Medicine*, *365*(11), 1033–1041.

5. Khoury, M. J., McBride, C. M., Schully, S. D., Ioannidis, J. P., Feero, W. G., Janssens, A. C., et al. (2009). Centers for Disease Control and Prevention. *Genetics in Medicine*, *11*(8), 559–567.
6. Marco, D. (2010). *Metagenomics: Theory, Methods, and Applications*. Caister Academic Press.
7. Oliver, F. (2023). In S. Santa, M. Timo, & N. Ana (Eds.), *Governing, Protecting, and Regulating the Future of Genome Editing: The Significance of ELSPI Perspectives*. Brill/Nijhoff.
8. Tobacco and Genetics Consortium. (2010). Genome-wide meta-analyses identify multiple loci associated with smoking behavior. *Nature Genetics*, *42*, 441–447.
9. Zhang, J. P., Lencz, T., & Malhotra, A. K. (2010). D2 receptor genetic variation and clinical response to antipsychotic drug treatment: A meta-analysis. *American Journal of Psychiatry*, *167*, 763–772.

Index

X

Y

Z

FSC
www.fsc.org
MIX
Papier | Fördert
gute Waldnutzung
FSC® C083411

Kolibri 360 GmbH
Ferdinand-Jühlke-Straße 7
99095 Erfurt, Deutschland
produktsicherheit@kolibri360.de